House of Invention

House of Invention

The Secret Life
of Everyday Products

DAVID LINDSAY

The Lyons Press

The essay "Muzak" originally appeared in
American Heritage of Invention & Technology
in somewhat altered form.

Design by Integrated Publishing Solutions

Photo on page 25 courtesy of the Schomburg Center
for Research in Black Culture

10 9 8 7 6 5 4 3
Library of Congress Cataloging-in-Publication Data

Lindsay, David, 1957–
House of invention : the secret life of everyday
products / David Lindsay.
p.   cm.
ISBN 1-55821-740-1
1. Inventions—History.   2. Inventors—History.
I. Title.
T19 .L56 1999
609—dc21   99-053253

For Gabe
The Can-Do Kid

Contents

Illustrations

Acknowledgments

T his book could not have been written without the help of many people, but special thanks must go to those who helped me through some especially lean times: Randy Wicker, Charles Bills, and my mother. Chris Pavone and Lilly Golden, my editors, deserve praise for playing the midwife with grace and skill, even during my more laborious moments. And, of course, I'm grateful to the inventors themselves for providing me with the raw material of their often difficult (yet ultimately fruitful) adventures.

Introduction:
Ghosts in the House

When I originally gave this book its title, I heard the emphasis falling on the word *invention,* but now I'm inclined to think otherwise. There are houses everywhere in these pages—the utopian houses imagined by Sir Francis Bacon, the shacks in which Madame Walker's "agents" buffed so many heads of hair, the magician's mansion with its electric bell at the gate, the "viewing household" brought about by the invention of television. All of these examples testify to the simple truth that a house is not the sum of its walls and ceilings and floors. Even the most sterile dwelling is alive with the ghosts of those who dreamed its contents into being. There are arguments here, and jokes, and mysteries.

The word *product* sums it up well. Any object, no matter how inert it appears, is invariably the result of a long and eminently human story. An invention may look like the paragon of convenience coming out of the box, but it was almost certainly born in crisis. In this sense, every invention is a form of communication, and the message it conveys is one of victory over disorder—not only the disorder of objects that fail to perform, but the whole spectrum of disasters, flukes, and oversized gestures we call the human folly as well.

Indeed, inventors can be said to resemble artists more than they do scientists. Like artists, they tend to float across categories with ease, and to make up their worldviews as they go along. With unerring frequency, their personal affairs go wildly astray, at great cost to themselves and those around them. That is a condition any musician or painter would recognize in a flash.

If this personality type—the inventor's personality—turns up so consistently in the archives, perhaps it's because human folly is intimately bound up with the actual work of inventing. Doing the "wrong thing," after all, is not so very different from doing the new thing. Put the other way around, if so many inventions start as mistakes, it's probably because some humans are especially prone to making them.

By the same token, the successful invention inevitably has a redemptive quality to it. In answering a technical problem, it stands as an offering to offset the many sins committed in its making. The invention, to the degree it makes life better, makes sense of a life oddly lived.

That said, inventors do differ from artists in at least one crucial respect. If by some miracle their ideas aren't stolen and the royalties should actually start rolling in, they can still count on languishing in obscurity. There are many reasons for this, not least that advertisers have their own interests in mind, but in any case the rule remains the same: to be an inventor—even one who changes the world—is to be largely unsung.

We have learned to contemplate ages past by the modern conveniences they did not possess, even to mark the advance of time by technological change. Yet somehow the creators and the consumers of these same technologies have been separated in the fray. If this book has one aim, it is to bring them together again . . . in the comfort of your own House of Invention.

House of Invention

1
THE BATHROOM

No matter what kind of house you live in,
the day invariably begins in the bathroom.
Behind closed doors, shaking off sleep,
you blindly reach out for help from
the inanimate world . . .

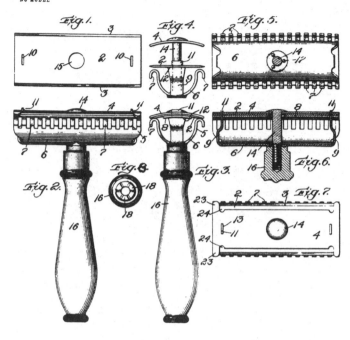

The Disposable Razor

King Camp Gillette is best known as the inventor of the disposable razor, but he was far more interested in reshaping civilization as we know it. These twin pursuits made life confusing for Gillette, and interesting for the rest of us. Indeed, as he ricocheted from pragmatism to principles, he managed to say as much about where America was headed as where it had been.

Gillette grew up in Chicago in the mid-nineteenth century, the son of a small businessman and part-time inventor. When his father lost everything in the great fire of 1871, King Camp landed a job as a traveling salesman for the Crown Cork & Seal Company, where the company's resident inventor, William Painter, was on hand to offer a crucial piece of advice. "Try to think of something like the Crown Cork," Painter told Gillette. "When once used, it is thrown away and the customer keeps coming back for more."

It's tempting to accuse Gillette of generating this anecdote after the fact. Certainly, his early career reveals no trace of a burning desire to invent, much less to make his name as a capitalist. On the contrary, in 1894, the thirty-nine-year-old salesman turned his back on bottle caps and began pouring his energies into a 150-page manifesto called *The Human Drift*.

Dedicated "To All Mankind," *The Human Drift* describes an architectural utopia by the name of Metropolis (coined several decades before Fritz Lang's movie of the same name), to be constructed at Niagara Falls. The choice of location was timely, to say the least. In 1894, the first large-scale alternating-current generators were being built at Niagara Falls, and unlike direct-current electricity, which had been the only choice just a few years earlier, AC could be transmitted over long distances—hundreds of miles even. No one knew exactly what the ramifications of such power might be, only that the word *power* was apt.

Gillette took his cue accordingly. His Metropolis, when it came to pass, would accommodate no fewer than one hundred million residents—at the time, the equivalent of the entire population of the United States. The central city was to span forty-five miles and comprise a series of ten thousand hivelike skyscrapers, each one 600 feet in diameter, with its own 250-foot-wide dining room. An automated distribution system would convey the fruits of what Gillette termed the "great Manufacturing Industries," themselves located in a set of concentric rings that included a congressional building, a circle of administration buildings, chemical laboratories, centers for experimental science and research, educational buildings, museums, and nurseries.

Climate-controlled interiors, vast transportation systems—once Gillette got going, there was no telling where he might stop. Yet, as much as the architecture of his Metropolis promised a bounty of futurism, its economic and political structure owed more to the early colonial charters of North America. Under the charter system, the colonies had been made up of "planters" who turned over their profits to investors for a number of years before gaining rights to the land. In Gillette's Metropolis, citizens who spent five years working for the "common good" would then be free to pursue a life of culture and leisure.

Gillette also borrowed his publicity techniques from the colonial era. In the early 1600s, investors often published glowing descriptions of their territories, regardless of the harsh realities, in hopes of drawing people away from the comforts of their homes. Eventually, this literary tradition inspired Sir Francis Bacon to write *New Atlantis,* which, with its depictions of sound machines and light shows and healing houses, amounted to a tourist brochure for a land that was, indeed, too good to be true.

The possibility remains that Bacon intended *New Atlantis* to be read as a satire, but that's another story. In Gillette's hands, the Baconian rhapsody attained new heights. "Would it not be a period of interest in the history of the world," he asked his readers, "that would make the blood race through one's veins with pleasure and excitement—a period in advance of all past periods and in advance of any period of the future, because it would mark a turning point in the history of man? To see this city rise like a beautiful picture sentient with life, reflecting the very essence of progress in its embodiment would make life worth living."

Readers, however, were satisfied with the country they lived in, and when *The Human Drift* sold poorly, Gillette was forced to reconsider his options.

In the nineteenth century, shaving was invariably accomplished with a straightedge razor, a time-consuming device that had to be stropped (sharpened on a leather strap) before each use. Gillette, remembering Painter's words, wondered if there might be a better way. What if there were a razor blade that could be thrown away once it had lost its edge? Of course, such a blade would require a new kind of razor, but this would only allow Gillette to design a safer razor while he was at it.

The Gillette system, patented in 1895, was just about as American as an invention could get—quicker, safer, easier—and when the American Razor Company started production in the fall of 1901, in a loft above a fish store on Boston's Atlantic Avenue, only one more obstacle remained. In order to keep costs down, Gillette needed a way to produce his blades on a large scale. Fortunately, John Joyce, a millionaire Boston brewer with $60,000 in capital, chose this moment to step onto the scene. With Joyce's backing (which he gave in return for a controlling interest), the American Razor Company was able to sell twenty thousand five-dollar razor sets in December 1904 alone. Gillette himself soon began to appear in national ads, making his image familiar in households nationwide.

Yet even in the glare of publicity, Gillette had not forsaken his dreams. In 1910, he offered Theodore Roosevelt, late of his post as leader of a nation, a job as president of Metropolis. (Roosevelt hastily declined the offer.) Eventually, Gillette's philosophical bent poisoned his business relationship with

Joyce so badly that he sold out his part of the razor business for close to $1 million and moved to California to write another book.

Among other things, *The People's Corporation* advocated relocating the entire world population to Texas, where it could be subsumed into a single corporate entity. No more nations, no more political leaders—everyone would belong to the same company, and everyone would own a share. Determined not to repeat the failure of his last book, Gillette then approached the socialist firebrand author Upton Sinclair for literary advice. If the first visit was unannounced, the rest were unavoidable. "Mr. Gillette was coming over two mornings a week to tell me his ideas," wrote Sinclair in his autobiography, "the same ideas over and over again."

Like *The Human Drift* before it, *The People's Corporation* failed both critically and commercially—and with it so failed the man. When Gillette died, on July 9, 1932, he was, by any ordinary standard, successful beyond his wildest dreams. Yet to the end he remained disappointed that his real dreams had come to naught.

Perhaps his disappointment was premature, though. Today, Americans are playing the stock market like never before, and there are moments when their holdings on Wall Street do seem to be converging into a single corporation. Gillette itself, now grown to multinational proportions, can take some credit for this trend, as can Warren Buffett, who built his fortune largely on Gillette and, in the process, inspired countless Americans to become day traders.

All that remains, it seems, is to convince the peoples of the world to move as one to the Lone Star State.

Vaseline

When he was going all guns, Robert Chesebrough put on a show worthy of Barnum himself. Standing before a rapt audience, he would burn his skin with acid, or sometimes, for variety's sake, with an open flame. He would cut his hands or arms with a knife or a razor blade. Having scared the bejeezus out of his audience, he would then spread a clear jelly over his injuries and display the scars from previous wounds—healed, he insisted, by this miracle product.

Was Chesebrough selling snake oil? Not exactly, but the curative powers of Vaseline were certainly bound up with an intoxicating promise that said a great deal about the culture in which it appeared.

In 1859, Chesebrough was a young chemist in Brooklyn, New York, who sold kerosene for a living. Americans had come to rely on the kerosene to light their homes, thanks to the great success of the whaling industry, which delivered the sperm oil from which kerosene was derived. By 1845, however, the overharvesting of whales had led to a shortage of sperm oil. Kerosene became scarce accordingly—and so did Chesebrough's income.

The petroleum industry began in earnest in 1859, when Edwin L. Drake drilled the first successful oil well, in Titusville, Pennsylvania. The early refineries produced about 75 percent kerosene, which could be sold for high profits. Needless to say, Titusville soon became a magnet for fortune seekers: by 1860, there were already fifteen refineries, or "tea kettle" stills, in operation.

Chesebrough must have had his nose to the wind, because he was among the first to arrive. Of course, as a mere chemist, he had no hope of drilling any wells of his own. Instead, he became intrigued with the gooey substance, known as rod wax, that stuck to the drilling rigs. The riggers hated rod wax for its tendency to make drilling rigs seize up, but had discovered that it was useful for healing cuts and bruises.

Though they probably didn't know it, the riggers of Pennsylvania were echoing the wisdom of the ages. American Indians had long recognized the medicinal value of petroleum (which appeared in small quantities in springlike formations, the

La Brea Tar Pits of California being a notable example), as had the ancient Persians and Sumatrans. Desert nomads had used it to treat their camels for mange. Charles V, the Holy Roman emperor, had found it a balm for his gout. Even in Chesebrough's day, petroleum was being advertised as a miracle cure for most anything. But until the invention of the oil well, petroleum had been too rare a commodity to justify any serious capitalist aspirations.

It didn't take long for Chesebrough to extract the key ingredient—the translucent material we know today as petroleum jelly. Marketing, however, was another matter. Convincing the public that a dab of smelly black sludge would be good for their health was going to take some doing.

Some of Chesebrough's marketing techniques were average enough. He combined the German word for *water* (*wasser*) with the Greek word for *olive oil* (*elaion*), to form the exotic-sounding *Vaseline*. He gave out free samples across New York State—in fact, he was among the first to employ the promotional giveaway—and within six months he had marshaled twelve horse-drawn buggies for distributing his product. But his "medicine show" took the Vaseline story to another level altogether.

In the years after the Civil War—the so-called

Gilded Age—figures such as John D. Rockefeller made enormous fortunes from the oil-refining industry. Oil itself was considered "black gold," a term that hinted of magical properties and, by extension, implied a certain alchemy in the transformation of ordinary citizens into millionaires. It was against this backdrop that Chesebrough appeared, cutting and stanching, a faith healer at work in the land of plenty.

To hear Chesebrough tell it, petroleum jelly could do just about anything, and by 1874 he was selling a jar a minute as customers tested the proposition. Vaseline replaced mustard plasters for chest colds, soothed chapped lips, and eased nasal congestion. It removed stains from furniture, polished wood surfaces, restored leather, and prevented rust. Druggists discovered that it made a good base for other medicines and ointments. Behind bedroom doors, it was enlisted as a sexual aid. In later years, the Chesebrough-Pond's Company put out a hair tonic for men (for that greasy look) and fishermen discovered its merits in attracting trout. In fact, the only use customers seem to have overlooked was the one Chesebrough proposed in the first place!

Then again, perhaps this was because the original claims weren't strictly true. When Vaseline was first introduced, Louis Pasteur—also in the habit of trying out medical products on himself—had yet to articulate the connection between bacteria and infection, and no one knew that petroleum jelly actually sealed cuts from further infection rather than healing them outright.

For his own part, Chesebrough remained magnificently undisturbed by the advent of germ theory. In middle age, he contracted pleurisy and saved himself from the grave—or so he claimed—by covering himself from head to toe with his wonder product. And in 1933, when he was in his nineties and truly near death, he revealed to the public that he had eaten a spoonful of the stuff daily for most of his adult life.

If Chesebrough exceeded himself at times, his unshakable belief continues to inspire devotion even today, albeit in unexpected corners. On April 19, 1998, Pastor Gary E. Yates gave a sermon at the Grace Baptist Church in Roanoke, Virginia, in which he described Chesebrough's flamboyant performances and then added: "People were lining up to buy Vaseline because the inventor backed up his belief with his behavior. The Thessalonians had that kind of faith. They didn't have a faith that you had to dust off to bring to church on Sunday morning. They didn't just have faith; they had an active faith."

It's a long way from snake oil to the sacrament, but if anything were ever qualified to make that leap, it's Vaseline—the nostrum that turned out to be useful after all.

Hair Straightener

Madame C. J. Walker was an unlikely candidate for fame and fortune. Born Sarah Breedlove McWilliams on the Louisiana Delta two years after the end of the Civil War, she grew up among sharecroppers who had never known anything but slavery. Her early life was provisional at best. She was orphaned in childhood and married by the age of fourteen. When her husband died, he left her with a daughter, A'Lelia. To make ends meet, she moved to Saint Louis, where she became a washerwoman at the age of twenty.

It was in 1905 that Walker began using her washtub to mix up pomades and ointments for straightening hair. Today, many black women straighten their hair strictly for style. In those days, however, the demands were different: as Walker saw it, her invention would help black women assimilate into white society.

How successful her customers were at assimilating is up for debate, but there's no doubting that Walker herself managed it. From hair straightener, she branched out into hair-growing tonics, soaps, shampoos, pomades, and various ointments, all of which she manufactured in her kitchen. For a time, she sold these items door-to-door in her neighborhood. Then she moved to Denver, where her brother lived, and started both a second marriage (to newspaperman Charles J. Walker) and a serious business in hair products. To a manufacturing plant in Denver, she was soon able to add another in Pittsburgh. By 1910, she had established a national headquarters in Indianapolis and was touting a cosmetic method known as the Walker Way.

Innovators had used the door-to-door sales techniques before (Vaseline inventor Robert Chesebrough among them), but Walker was the first to show that it could be adapted specifically to women. Long before the Avon Lady appeared in America, Walker was dressing her employees in a trademark outfit and sending them off to make house calls. When the lady of the household answered the door, she would behold a "beauty culturist" in a long black skirt and starched white blouse brandishing a bewildering array of gadgets—a hair-straightening comb, hot-iron curlers, hair softeners, and other ointments, all produced from a single, economical black satchel. If the customer took the bait, the beauty culturist would step in and dress her hair in the comfort of her own home.

Once it got started, the Walker empire grew by leaps and bounds. Walker personally trained every one of her agents, and ultimately employed as many as three thousand of them in the United States and the Caribbean. At a time when the public life was conducted largely in person, she went on lecture tours and demonstrated her wares tirelessly. On a trip to Paris, she caught the attention of the cause célèbre Josephine Baker, who agreed to have her hair done the Walker Way. When a French company copied the result, the "Baker Fix" became the hit of Paris. In the end, Walker owned a nationwide chain of beauty schools and an eminently famous face, thanks to her policy of adorning her products

with a photograph of herself. No believer in stock, she was the sole owner of her company.

All this industry eventually paid off: not only was Walker the first black woman to earn more than a million dollars, she was the first woman millionaire, period. But while her fortune allowed her to enjoy the finer things in life—she commissioned the black architect Vertner Tandy to build a mansion, at a cost of a quarter of a million dollars—she never forgot her origins, either. She established scholarships at the renowned Tuskegee Institute, where literary luminaries Ralph Ellison and Albert Murray later got their schooling, and throughout her life gave generously to the NAACP. Infected by her philanthropic spirit, Walker's employees took to forming Walker Clubs, which devoted themselves to community service.

When Walker died, in 1919, A'Lelia followed in her footsteps, managing a large trust fund for black orphanages, old-age homes, and schools, in addition to running the family business and establishing a literary salon called the Dark Tower. A'Lelia never followed her mother to the Patent Office, though; the successor on that count was a woman named Marjorie Joyner.

A native of Mississippi, Joyner moved to Chicago and began working for the Walker company at the age of twenty. After a time, she noticed that the Walker Way, for all its many advantages, was lacking in one significant detail: once a woman's hair had been dressed, there was nothing to keep it that way. In 1928, she put an end to this problem by patenting one of the first permanent-wave machines, which not only kept the Baker Fix fixed but simplified some of Walker's other techniques as well.

Although Joyner received no profits from her invention, she remained faithful to the company to the end, eventually rising to the position of national supervisor. For a time, she worked for the Works Progress Administration and, in 1945, she co-founded the United Beauty School Owners, a black-only beauty school, with McCloud Bethune.

Like Walker before her, Joyner advocated grooming as a means of advancement. Then she went a step further and urged the dressing of hair for men and women alike. Her urgings were certainly heeded: the practice of "konking" was nigh on universal in the black community until the 1960s.

After Malcolm X described in his autobiography how he was forced to thrust his head into a toilet to get the burning lye off his scalp, many blacks were quick to embrace the more natural Afro. Today, however, hair styling has come back with a vengeance—and this time among people of all colors. Indeed, Madame Walker's legacy has made it almost impossible to say who's assimilating into what—and that, perhaps, is her greatest success.

2
THE KITCHEN

The kitchen, some say, is the temple of the house—the site where lower life-forms are sacrificed to the human endeavor. Of course, the inventors of kitchen conveniences made some sacrifices of their own . . .

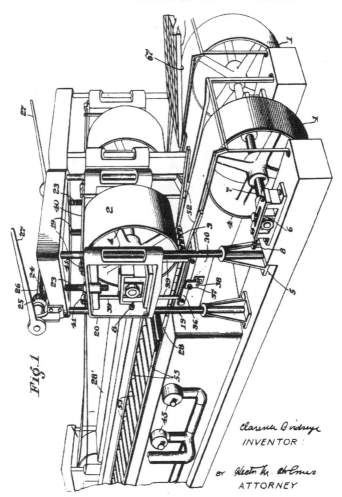

Fig.1

Clarence Birdseye
INVENTOR:

BY Hector M. Holmes
ATTORNEY

Frozen Food

Clarence Birdseye was fresh out of college in the early 1920s when he took a job as a naturalist for the U.S. government and went on an expedition to Labrador. Observing nature in such places was quite dangerous at the time; around the same period, a scouting party in Labrador had gone astray and frozen to death. Given such harsh conditions, it's reasonable to argue that Birdseye, in traipsing off to northern Canada, was equal parts naturalist and explorer.

In fact, in a strange sort of way, Birdseye was the *last* of the explorers. The same outward thrust that had taken Europeans around the Cape of Good Hope in the 1500s had also taken them to Canada, and in both cases the goal had been the same. They had been looking for the fastest route to the spices of the East, which held great value not only for their taste but for their ability to preserve food. Four centuries later, Birdseye ended up in the business of food preservation, too. It's just that he went about it in a different way.

Watching the Native Americans of Labrador as they fished, Birdseye noticed that the fish froze almost completely as soon as they were caught. More importantly, after they were cooked, they were scarcely different in taste and texture than fresh fish were. The key, Birdseye realized, was speed: slow freezing resulted in the formation of ice crystals, which destroyed the cellular structure of the fish. But if the fish were frozen quickly enough, the crystals had no time to form and the cellular structure was preserved. Having been a biology major at Amherst, Birdseye also surmised that freezing would preserve the nutrients—an educated guess that turned out to be correct.

In the early 1900s, many people were experimenting with mechanical and chemical methods to preserve food. Birdseye (who had already invented a stage light) realized he had just stumbled upon the best method of all.

Returning to New York, he lost no time in capitalizing on it, either. In 1924, he opened with a double punch: he founded Birdseye Seafoods, Inc., and, at the same time, developed a process called "quick-freezing," which packed fresh fish into food cartons, then pressed them between two refrigerated surfaces. From there, he started in on vegetables. In 1929, he accelerated the freezing process with double-band freezing, which chilled food along two different planes at once. Knowing how to count as well as the next guy, he ultimately developed a process known as multiple-plate freezing. (This is essentially the same process in use today.) His pièce de résistance, perfected in 1930, was a system that could pack dressed fish, meat, or vegetables into waxed-cardboard cartons, then flash-freeze them under high pressure (patent no. 1,773,079)— a kind of "one-stop" freezing machine.

Having developed his product, Birdseye trained his sights on distribution, no simple matter when it came to a technology born in the Arctic. In fact, Birdseye's distribution dilemma revisited a problem encountered by Thomas Edison when he invented his lightbulb. A lightbulb meant nothing by itself in the days when only crude batteries existed, so Edison had been compelled to invent a complete system for delivering electricity into people's homes.

The same went for Birdseye: once his foods were frozen, they had to stay frozen. To this end, he began testing refrigerated grocery display cases in 1930 and entered into a joint venture to manufacture them four years later. By then, of course, the nation had been plunged into the Depression, and it wasn't until 1944 that his company was able to lease refrigerated boxcars. But by the time of his death, in 1956, national distribution had become a reality and frozen foods were beginning to be a familiar sight in supermarkets around the nation.

The frozen-food industry has only grown since then. In 1994, Global Market Research reported total retail sales of frozen foods to be $9.8 billion. On the shelves of supermarkets across America, one can find frozen pizza, frozen dinners, and frozen soft pretzels; frozen ravioli, frozen peas, and frozen broccoli. In the Shandong Province of China, there are even "frozen food towns."

These towns, which are run by Katokichi Co., Ltd., give new meaning to the term *prodigious*. According to the director of the overseas business division, Mr. Saburo Sasaki, Katokichi has established ten production bases in Shandong Province in the past three years under its "frozen food town" plan. A relatively new factory in Weihai City is said to produce thirty tons of frozen food a day—among the largest output of any food factories in China. In order to handle such huge masses of food, the company has been forced to lean heavily not only on Birdseye's freezing methods but on his accomplishments in distribution, too; Katokichi currently exports much of its food to Japan, by way of the ports of Qingdao and Yantai.

Frozen foods represent the tail end of a progression that began with the spice trade and moved on through canning and refrigerated boxcars, until at last the culinary arts have been completely unyoked from the seasons. We can eat peas in February and pumpkins in June. This again brings to mind a parallel to the electric light. Both Edison's lightbulb and Birdseye's flash-freezer have separated us from the natural rhythms of life—one from the whirl of each passing day, the other from the flow of the seasons.

It's hard to argue with the Birdseye copywriters when they claim that "Clarence Birdseye has indirectly improved both the health and convenience of virtually everyone in the industrialized world." Nevertheless, in recent years, the increasingly "abstract" diet brought about by frozen foods has come under attack. This is most obvious in the case of macrobiotics.

A philosophy popularized in America by Michio Kushi, macrobiotics, among other things, emphasizes a seasonal, local diet. Thus, the true macrobiotic living in New Jersey will have a diet rich in tomatoes and turkeys, while his Floridian peers will subsist on yellowtail snapper and oranges.

Macrobiotics also comes with a strong moral component in tow: to eat outside your own time and space, it says, is to lose the spiritual essence of the food.

In this respect, it's worth considering the eating habits of the Native Americans from whom Birdseye learned his lucrative technique in the first place. The tribe Birdseye encountered was most likely the Metis, descendants of the Inuit, who have lived in Labrador and Newfoundland for centuries. Until recently, the Metis were much like macrobiotics, in that they ate what was in season. In fact, they organized much of their lives around hunting—trapping in autumn, hunting seals, caribou, and game birds in the spring—always with deference to the will of what they called the "animal masters." In a nutshell, the animal masters required contrition for killing, ritual treatment of the carcass, and—most relevant to our case—a vow not to overharvest. If one brought home too many carcasses at once, it was an invitation to bad luck in the next hunting season.

Of course, the ability to freeze food is practically a mandate to overharvest, as evidenced by Katokichi's admission that the great success of its frozen food towns has forced the export of its surpluses to

Japan. As a species, we are overharvesting like never before, in large part because we can.

Even the macrobiotics crowd has failed conspicuously in this area. Rather than eating foods grown locally, it has long settled into a diet of foods grown locally *in Japan,* with rice, tofu, and tempeh ranking high on the list. Doing that absurdity one better, Jaclyn's Food Products of Cherry Hill, New Jersey, currently offers dinners that are both frozen *and* macrobiotic. If no animal masters have been hurt in the preparation, these meals suggest nonetheless that Birdseye left an important strain of Inuit wisdom behind when he trudged home from the wilds of Labrador.

The Blender

Mediocre entertainment has often inspired exceptional technology. One need look no further than the many dull movies with fantastic special effects to see that this is so, but there are examples from the more venerable section of invention history as well. Alexander Graham Bell, for one, was considering a career as a playwright until his brother bemoaned "the extreme poverty of thrilling ideas in your dramatic works" and prompted him to concentrate on the telephone instead. A more direct example of the "mediocrity principle" at work is the blender, which was born, as it were, in the trunk of a journeyman musician.

The earliest version of a blender was invented in 1922, by Stephen J. Poplawski of Racine, Wisconsin. Poplawski had been working on various methods of mixing beverages since 1915. Unfortunately, he failed to imagine the possibility of grinding up solids into a viscous concoction. His patent detailed "an agitating element mounted in the bottom of a cup and a driving motor mounted in the base," but it was designed to mix only liquids. No doubt in this he was influenced by the interests of his employers at Arnold Electric Company, who were focused almost entirely on the exploding soda-fountain industry.

Another Racine native, Frederick J. Osius, was thinking along the same lines in 1933, when he patented his first "drink mixer." Then, somewhere around 1936, Osius got the idea for a "disintegrating mixer for producing fluent substances." This was clearly an improvement over the simple drink mixer, but it was still in the design stage. Osius needed a way to bring his disintegrating mixer to perfection, and from there to market.

As it turned out, one of Osius's investors was the brother-in-law of the publicist for Fred Waring, a well-known big band leader of the day. It seemed reasonable enough to approach Waring as a potential promoter, and the opportunity presented itself when Waring was playing a radio performance at the Vanderbilt Theater in Manhattan. Osius himself made the call. Dressed in striped pants and a lemon yellow tie, he finagled his way backstage after the show by claiming to have an appointment. There he found Waring and made both pitch and demonstration on the spot. The machine did not work, but Waring was interested anyway.

Waring's band was not among the greatest musical outfits ever—its arrangements tended toward the sweet and mushy. Ironically, these qualities were perfectly suited to the machine at hand. As Waring himself later put it, saying as much about his musical tastes as anything, "I had been wanting a machine to make absolutely velvet-like, sweetest-possible banana milk shakes all my life."

Velvetlike, sweetest possibilities apparently made all the difference: where Poplawski had simply patented his conception of a blender, Osius was buoyed along by the dynamics of Waring's popularity. This popularity came with money in tow, of course, and after six month's time Osius had spent $25,000, courtesy of Waring, on research and development. But still the machine didn't work.

The basic problem was that the drive shaft had to go through the bottom of the container, which meant that any liquids in the container tended to leak. Seeing that things had reached an impasse, Waring hired engineer Ed Lee (Waring himself had studied engineering at Penn State) to devise the right combination of leak-proof seals and couplings at the base of the container. Then the bandleader turned his attention to cosmetic changes, hiring Peter Muller-Munk, a German designer, who gave the machine its distinctive, chrome-and-glass, Art Deco look. Waring himself provided the "o" in *Blendor*, and they were in business. (Patent no. 2,109,501, granted to Osius in 1938, shows the Art Deco design fully blown.)

The Waring Blendor debuted at the 1937 National Restaurant Show in Chicago, but it came to life in the realm of entertainment, as Waring took every opportunity to tout its benefits. He loved to whip up bizarre frappés backstage and dare his musicians to down them. (The musicians reportedly became expert at finding ways of evading this responsibility.) He had a theatrical trunk designed to order that functioned as a Blendor demonstration kit. And, of course, he used radio to promulgate the wonders of the Blendor far and wide. The year 1938 found him acting something like an acoustic cooking-show host on WEAF Radio as he described the making of a banana-chocolate milk shake on the air: "At the bottom of this jar there's a little revolving band-shaped thing like a propeller [spinning at] twelve thousand revolutions a minute."

The Depression had already slowed the sales of the Waring Blendor (its original price was $29.75—a week's pay for some back then), and World War II only made matters worse, so in 1947 Waring sold his company to an entrepreneur named Hazard Reeves. Until that time, the Blendor had been sold exclusively to bars and restaurants; Reeves foresaw a wider market, and began selling it as a household product. Then the invention made an unexpected leap into the world of medicine.

Hospital workers quickly discovered that Blendors were ideal for churning out baby food. The next step was the Waring Aseptical Dispersal Blendor, which Jonas Salk allegedly used in the 1950s for grinding up materials in his pursuit of the polio vaccine. (The centrifuge, a medical device used for slightly different purposes, was developed by Major Harry Armstrong and Dr. John Heim in 1636, while they were working on aviation medicine for the air force.)

Indeed, Waring Blendors are still used for medical purposes today. In a 1993–94 paper titled "The Chemistry of Proteins," Dr. Louis Kanarek and a team of researchers advocate a Waring Blendor over other available methods for "the purification of fimbrial lectins from different *E. coli* strains, as well as from *Klebsiella* and *Serratia* species."

Not bad for an invention that began as a convenience for soda jerks.

Breakfast Cereal

Whenever you pour yourself a bowl of Kellogg's Corn Flakes, you're not just pouring yourself a bowl of Kellogg's Corn Flakes. You're also getting mixed up in a contest between two brothers who've been dead for years.

In some ways, the Kellogg brothers were typical siblings. The older of the two, Dr. John Harvey Kellogg, was domineering, charismatic, and intensely driven to succeed. Will Keith Kellogg, on the other hand, was an expert accommodator who constantly sought the approval of his demanding older brother. This dynamic will come as no surprise to birth-order theorists. What was different about the Kelloggs was the outsized nature of the life they led.

In 1876, John Harvey was twenty-four years old and fresh from a stint at New York's Bellevue Hospital when he arrived in Battle Creek, Michigan, to take a position as physician-in-residence at a sanatorium run by the Seventh-Day Adventist Church. The religious group had already instituted a regimen of vegetarianism and hydrotherapy to aid its ailing patients, of which there were exactly twelve at the time. John Harvey would expand enormously on their methods and, in the process, swell the living quarters well past capacity.

As a first step, John Harvey changed the name of the place from *sanatorium* to *sanitarium*—a word of his own minting that served no apparent semantic purpose except to announce his arrival at Battle Creek. Then he began working in earnest. He devised dinner menus that broke down the nutritional content of each dish into carbohydrates, proteins, and fats—an innovation that has survived quite tenaciously down to today. He instituted the practice of counting calories. A staunch man of temperance, he forbade smoking and drinking anywhere on the grounds. And, of course, he also continued the policy of the Seventh-Day Adventists, banning meat from his patients' diets. (Preventing the opening of a "meat speakeasy" across the street proved more difficult. Even Duke, John Harvey's famously vegetarian Saint Bernard, was able to get bones on the sly from patrons of the Red Onion.)

In other words, John Harvey Kellogg was your average health-food enthusiast who just happened to have been born several generations before his time. But he was also a little weird. Aspiring to something like a composer working in the medium of wetness and heat, he developed more than two hundred different kinds of baths, douches, and "fomentations" (read *compresses*), in the belief that "hot and cold water accomplish a wide variety of results, in single or in a large number of complications." He used electric-light baths, diathermy, and thermopenetration. He massaged his patients with vibrations, X-rayed them and irradiated them, and all but rearranged their internal organs.

It's possible that this great display of solutions stemmed from a belief that time was running out. John Harvey predicted that a time would come when mental illness would increase beyond the ability of medical institutions to house the sufferers. Implicit in this statement was the belief that his own regimen, which went by the label of Biologic Living, would prevent the epidemic of madness from ever taking place. *Maybe.*

Certainly, John Harvey practiced what he preached. On a typical day, he would rise at four in the morning, skipping breakfast in order to perform his first surgical operations of the day. By eleven he would be ready for the daily board meeting, which often lasted until late afternoon. Then it was time for conferences with patients, writing (he was forever writing books), experiments on the gastric juices of dogs, refining his hydrotherapeutic machinery, and so on. Sundown had no effect on his energies whatsoever. To dictate for eight hours at a stretch was nothing to him. His personal best was twenty-four hours of unbroken dictation.

Battle Creek was a veritable beehive of activities directed toward John Harvey's singular vision. But John Harvey also produced more work than he could handle, and eventually he had to call on his brother to keep the sanitarium from collapsing under the weight of his genius.

In the early 1880s, Will Keith Kellogg left a string of dead-end jobs—including those of a stable boy and a broom salesman—to work for his brother at Battle Creek, and he remained there for twenty-five years. He never held a specific title. Indeed, it would have been hard to formulate one. His obligations ranged from the pursuit of escaped mental patients to the operating of the magic lantern for his brother's customary Monday night lectures. By 1890, John Harvey's mail amounted from sixty to one hundred letters a day, each of which required a reply from W.K. No one considered it bizarre that W.K. worked 120 hours between Monday and Friday and then, on Saturday, arranged for ambulances to meet new patients at the train station. "If you want anything done," one doctor is said to have advised, "go to W.K. He will listen to your story and he will give you an answer and the answer will be perfectly fair, and it will be accomplished as he says!"

And yet, swamped as W.K. was, his dashing older brother still felt entitled when they quarreled to accuse him of being "a loafer."

The tension between the two brothers makes it almost touching that they invented Battle Creek's most enduring legacy together. At the turn of the century, the typical American breakfast consisted of meat, eggs, fried potatoes, boiled coffee, bread slathered in molasses, fried mush or cornmeal, and pie for dessert. Such a punishing diet might have been all right for heavy laborers, but the increase in sedentary jobs was changing things, and the medical industry was beginning to notice as much.

John Harvey himself had already been influenced by Dr. Sylvester Graham, inventor of the Graham cracker: as a student at Bellevue, he had actually subsisted for a time on a diet of Graham crackers and apples. (This interest in new forms of food survived the move to Michigan and eventually found expression in the Sanitas Food Company—a spinoff from the sanitarium.) One day, a patient brought a sample of shredded wheat to Battle Creek, which led John Harvey to seek out its inventor, a Denver attorney named Henry D. Perky. When the meeting failed to materialize, John Harvey resolved to make a better food himself.

So it was that in 1894 John Harvey and W.K. began tinkering around with the idea of breakfast cereals. Years later, John Harvey would claim to have invented corn flakes in a dream, but the truth was actually less lofty than that. The experiment began with wheat, which W.K. boiled and then gave to John Harvey, who took it to the basement and fed it through a set of rollers that were normally used to grind granola (which was itself a Battle Creek creation). While John Harvey was thus employed, W.K. squatted—literally beneath his brother—and scraped the sticky, gummy dough off the rollers with a chisel.

In the end, it wasn't any particular stroke of brilliance as much as their own monstrous workload that turned the trick. The Kelloggs had prepared a batch of wheat to be run through the rollers, only to be waylaid by other duties. One or two days later, when they were finally able to return to the project, they found that the wheat had grown moldy. After some discussion, they decided to run it through the rollers anyway, and much to their amazement, each wheat berry came out as a perfectly compressed flake. When they were baked in a nearby oven, the flakes tasted moldy, but other than that, they were not bad at all.

From there it was a simple matter of trial and error before they discovered the secret. In the process of molding, the moisture had been spread evenly throughout the wheat. The same effect, as they eventually discovered, could be achieved without the moldiness—by letting the cooked wheat stand for several hours in a tin-lined bin. What's more, the process worked just as well on corn as it did on wheat. They worked up a sample batch, and the sanitarium patients approved. (The success of corn flakes came later, after W.K. found a way to use corn grits instead of the tougher corn kernels.)

Characteristically, it was John Harvey who applied for the patent on flakes made from "wheat, barley, oats, corn and other grains," and it was he who took the credit in public. It was the younger brother, however, who turned their invention into an industry. John Harvey, though a tireless promoter of his scientific beliefs, frowned on large-scale advertising campaigns and commercialism in general, and consequently he did as little as possible to help when W.K. decided to turn corn flakes into a business.

The last straw came in 1905, when W.K. built a factory at a cost of $50,000 and John Harvey refused to reach into the sanitarium coffers to pay for it. The following year, the two brothers severed ties completely. W.K. went on to devise a series of brilliant advertising campaigns—including one that offered a free box of Kellogg's Corn Flakes to any woman who winked at her grocer—and watched his empire grow. John Harvey, meanwhile, continued on with his plans for Biologic Living, which gradually faded from fashion.

The split was among the most severe in the annals of business. Only after the passage of many years did W.K. feel compelled to write, "In our food business the Doctor and I were doing business together. Some of the formulae he worked out, some I did, and he made suggestions and I made suggestions, and I think he took most of the credit for the work I did."

Even so, W.K.'s sense of inferiority never quite left him. As late as 1939, when he stood at the top of the Kellogg empire and John Harvey had been almost forgotten, he was still deferring to his older brother. To his advertisers, he wrote, "The advertising should not imply that I claimed full credit for the invention for, if it does, I am afraid Dr. Kellogg's feelings would be injured."

It is interesting to consider that today Dr. John Harvey Kellogg is remembered mostly for his eccentricities (as caricatured only slightly in T. Coraghessan Boyle's *Road to Wellness*), while the signature of Will Keith Kellogg has gone on to grace a hundred million corn-flake boxes. It is also worth pondering the younger brother's motives for taking part in the corn-flake venture in the first place. John Harvey often skipped breakfast in his hurry to skip lunch and dinner, and a breakfast cereal—at the time a novel idea—clearly offered him a way to take his morning meal without having to interrupt his highly important work.

Was Will Keith simply trying to make sure that his brother was fed? And by extension, has the Kellogg company been doing the same for its customers ever since?

No time to answer: you're late for work.

3
THE FOYER

Getting out of the house is rarely a simple matter. Where have you left your keys this time? Your wallet? If it's any consolation, the people who came up with the things you need to get through the door didn't have it so easy, either . . .

The Intercom

A ccording to some accounts, the intercom was invented in 1920 in Berlin, where doctors called it a "door telephone" and used it for night calls. Another version gives the honors to Joshua Cowen, inventor of the famous Lionel trains, who presented plans for an electric doorbell to a teacher in the 1890s, only to be waved away as a hopeless dreamer. Yet long before either of these events, the great French magician Robert-Houdin had already installed a fantastically complicated intercom in his country home. In fact, Robert-Houdin elaborated the device so completely as to create a prototype for today's much predicted automated house.

Robert-Houdin was born Jean-Eugène Robert in Blois, France, on December 6, 1805. After a half-hearted stab at a medical degree, the young Jean followed in his father's footsteps and became a watchmaker. Marriage came in 1830, to Josephe

Cécile-Houdin, and a move to Paris followed soon after. In the capital, the newly named Robert-Houdin made the acquaintances of the magician Comte and the ventriloquist Roujol—his first forays into a life beyond repairing watches.

Then, during a protracted illness, Robert-Houdin decided—somewhat counterintuitively—to keep his finances afloat by constructing an automaton that could write intelligible words on a page. In 1844, this mechanical penman won a prize at the Paris Exhibition, and he cleaved to his counterintuitions ever after. The following year, he made his debut as a stage magician, with a "temple of mystery" room at 164 Jardins de Valois in the Palais Royal.

Although often touted as a great originator, and sometimes as the greatest magician of all time, Robert-Houdin actually learned many of his tricks from others. Still, one fact cannot be disputed: he understood the dramatic potential of electricity (which the scientists themselves were at pains to explain) better than anyone else in his time.

One of his favorite tricks was to present a small object to an audience and then, placing it near the edge of the stage, to ask a volunteer from the house to lift it. Before the start of the show, Robert-Houdin had placed an electromagnet under the floorboards, which held the metallic object on the floor no matter how hard the volunteer strained and grunted. Of course, when a switch was thrown, the magician expressed his confusion and lifted the object with ease.

If this ruse impressed Europeans, who had at least heard of electricity, it positively flabbergasted Algerians when the French government sent Robert-Houdin to their country to put down a rebellion by exposing their holy men as tricksters. But it wasn't until he retired to his country home in St.-Gervais, near Blois, that his electrical talents reached their apex.

Not surprisingly, Robert-Houdin arrived at his first doorbell by way of an illusion. Calling his children to his side, he told them, "Here is my new trick. When I put my finger in this tumbler of water, Adèle will enter the room!" Unbeknownst to the children, he had fixed a simple electric switch beneath his foot with a wire leading to a bell in a separate room, where Adèle, the servant, awaited her call.

Before long, this device metamorphosed into a doorbell proper. Whenever a visitor reached the entrance gate at St.-Gervais and lifted the knocker, a loud ringing immediately became audible in the house, which stood about a quarter of a mile from the gate. The ringing continued until a servant pressed a button on the wall, at which time the gate unlocked and an enameled plate appeared, bearing the command "Entrez."

Not content merely to mystify his guests, Robert-Houdin eventually took to monitoring them as well. As he arranged it, the gate closed by a spring, the opening and closing of which set another bell ringing in a kind of Morse code. The peculiar rhythm of these rings told the master of the house how many visitors had arrived, and whether they were first-time callers or friends of the family. (Presumably, friends knew how to send a secret signal.) A separate system of bells not only told Robert-Houdin when the postman arrived and how many letters he had brought with him but told the postman in turn whether a package was waiting for him at the house. (Robert-Houdin wrote all of his letters at night and didn't care to be roused when the postman came.)

Robert-Houdin was not averse to a little manipulation if the circumstances called for it. When he learned that the groom of his favorite horse, Fanny, was spending the money meant for oats on himself, Robert-Houdin rigged up a clock in the house that parceled out the oats from afar. As a former watchmaker, he was, of course, a natural with timepieces. In his bell tower, he installed a *self-winding* clock that was powered by the comings and goings of servants through a swing door in the kitchen. Just for kicks, he rigged this clock so that he could slow it down or speed it up, depending on when he wanted to take his meals. Needless to say, the cook, who had been hired for his punctuality, did his share of double takes.

Robert-Houdin's estate became something of a legend in France and gave him the reputation of being a sorcerer. Still, all the powers at his command were not enough to keep tragedy from striking. In a letter of September 11, 1870, the magician copied verbatim a soldier's account of the death of his youngest son, Captain Eugène Robert-Houdin:

Upon the order of Capt. Robert-Houdin, Lieut. Girard advanced with two men to reconnoitre the enemy. He took three steps and fell, crying: "Do not give up the Coucou" [a familiar expression applied to the flag]. We carried him away and the Captain shouted "FIRE!"

The order to retreat came, but we did not hear it, and continued to beat against a wall of fire . . . Soon our captain fell, saying: "Tell them . . . that I fell facing the enemy." A bullet had pierced his breast. He was taken in the ambulance to Reichshoffen where he died, four days later, from his wounds.

The order to retreat came, but they did not hear it. One can only imagine what Robert-Houdin, whose expertise had once quelled an insurrection, would have devised in the way of battlefield signals had he been able. As it was, he died not long after, on June 13, 1871.

The house in St.-Gervais was immortalized several decades later when Raymond Roussel reimagined it in his classic Surrealist work *Locus Solus,* but the publication of this book sparked no sudden craze for building crazy homes. In years since, the doorbell and the intercom have dwindled to their most functional forms, while Robert-Houdin's more whimsical contraptions have expanded into a sobering array of security cameras and alarms. The magician's mansion, meanwhile, lives on in the dream of the fully automated house, which, like a trick of another sort, never quite seems to arrive.

Bank Notes

As with many inventions, paper money was originally the brainchild of the Chinese. The European tradition began in Venice around 1587, when bankers at medieval fairs began to use endorsements as a way of transferring the receipts of depositors. The first bank notes proper, by some accounts, were issued in 1658 by the Riksbank of Stockholm. But none of these examples constituted paper money as we understand it today—that is, as a permanent currency negotiable on the international market. Before that could come to be, a seafaring rogue had to hold a position of high esteem in colonial New England.

William Phips was born on a farm in Kennebec, Maine, in 1651, the last of (count 'em) twenty-six children. Rather than attend school, he learned carpentry and set off for Boston, where he met a ship captain, Roger Spencer. By marrying Spencer's widowed daughter Mary, Phips was able to become captain of a trading sloop, and, like many mariners of his day, he was soon sailing to the West Indies in search of profits.

At the time, the Caribbean was rife with stories of sunken treasure, and several old tars provided Phips with some sound information on a fleet of sixteen Spanish galleons that had gone down laden with plunder. The Spaniards, Phips was told, had sailed from Puerto Plata in 1643 and were heading north through treacherous Bahamian waters when a hurricane sent the expedition to its watery doom.

Phips calculated where the fleet had sunk, then set off for England to find backers for his expedition. Somehow, he managed to catch the ear of King Charles II, who put him in command of a Royal Navy frigate. Phips sailed from England in 1684, but hot weather and a recalcitrant crew waylaid the expedition. It took another three years and another round of pitches to English noblemen before he could set out for the Caribbean again and find his prize.

In 1688, Phips returned to London, his ships weighed down with 37,538 pounds of pieces of eight, 25 pounds of gold, and 2,755 pounds of silver, one-sixteenth of which was rightfully his. Rich beyond any reasonable calculation, he was knighted and, eventually, appointed the first royal governor of Massachusetts.

None of this would have anything to do with the appearance of paper money, perhaps, except for the matter of Phips's personality. In 1692, when he became royal governor, Phips established himself as a crude braggart and something of a crook, and for a time embarked on a flagrant kickback scheme: whenever a vessel was condemned, he would sell it for a pittance to a friend, who offered it on the open market at full price, with eleven-sixteenths of the proceeds reverting to Phips. (No doubt it was antics like this that inspired Phips, at one point during his tenure, to cane the customs collector in public.) Indeed, the only thing he knew how to do well was to look for booty. And when none was to be found, another solution had to be forged.

In 1690, still fresh from his Caribbean success, Phips laid siege to the Acadian capital of Port Royal and, since the French garrison consisted of only some seventy men, he won without a fight. He spent twelve days pillaging Port Royal, then had his way with the rest of Acadia: Castine, La Havre, Chedabucto, the settlements at the head of the Bay of Fundy.

No sooner had Phips returned to Boston in a triumphant flourish than he began mounting an attack on Quebec. Unfortunately, he neglected to consider several crucial factors this time. For one thing, his fleet was leaving Boston just as the harsh Canadian winter was approaching. More important, his crew of some two thousand men had been promised their reward in plunder—the standard arrangement for privateering missions—at a time when the colony's treasury had all but run dry.

Phips sailed for Quebec on August 20, 1690, and by October 23 he had made such a vague impression on the French that he was forced to head back to Boston empty-handed. According to the New England historian Thomas Hutchinson, "The government was utterly unprepared for the return of the forces. They seem to have presumed, not only upon success, but upon the enemy's treasure to bear the charge of the expedition. The soldiers were upon the point of mutiny for want of their wages. It was utterly impracticable to raise, in a few days such a sum of money as would be necessary."

A few Massachusetts merchants did stand forward in November 1690 to put up an emergency loan should the expedition fall shy of its goals. But when Phips arrived in Boston on November 18 with a full account of the disaster, it was clear something had to be done at once. (It seems never to have occurred to Phips, a man possessed of vast quantities of Spanish gold, that he might have bankrolled the expedition himself.)

New Englanders had actually considered the use of paper money before. Forty-four years earlier, the idea had been bandied about for some six months. In 1681, a more serious attempt had been made, with actual bills being issued "in the nature of a money bank or merchandize lumber to pass credit upon, by book entries; or bills of exchange, for great payments and change-bills for running cash." Lest anyone confuse the matter, it was also stated that "credit pass'd in the Fund, by book, and Bills, (as afore) will fully supply the defect of money." But these ventures hardly armed the Massachusetts General Court with seasoned experience in negotiable bills, and the experiments in Sweden in 1658 might as well have never happened.

And so the General Court took a deep breath and issued forty thousand British pounds' worth of paper bills, known as old colony or charter bills, which could be turned in to the treasury in lieu of taxes. (In one of the more interesting gestures in the history of graphic design, the designer of the twenty-shilling note saw fit to place the image of an Indian inside the seal, despite the fact that the colony was at war with the Indians at the time!)

The experiment got off to a shaky start. Some claimed that the old colony bills that went into the treasury never came out again. The soldiers who had originally been targeted as recipients complained that the bills were valued at twelve or thirteen shillings to the pound, forcing them to purchase useless items just to be rid of the paper. With the treasury still under par, the time between issuance and redemption by taxes had to be increased again and again. If all this weren't enough, the very fact of paper money only deepened anxieties at a time when the Salem witch trials, war with the French, enforced freedom of religion, and the first attempt at a newspaper were gripping the colony.

But the die had already been cast. By 1694, London had followed suit and set up the Bank of England, influenced in part by the Massachusetts experiment. By 1715, Massachusetts itself, which had been thinking purely in terms of emergency measures, found that the old colony bills had driven silver out of the region and taken pride of place by default.

As for Phips, he was called back to England in the fall of 1694 to answer for his imperious ways, but fate threw one last favor his way: on February 18, 1695, he died suddenly, before any formal proceedings could begin. He was buried in London, in the yard of the Church of Woolnoth—the man whose drive for solid gold forced the development of paper money.

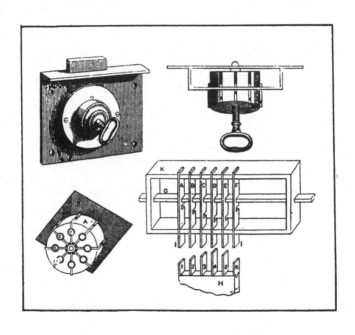

Locks and Keys

When today's hackers break into top-secret computer systems and announce their victories to crestfallen CEOs, they're following in a very old tradition. Long before cryptology became a question of national security, lock pickers were gleefully outwitting the best efforts of locksmiths to keep them out. And one of them could not be stopped for love or money.

Locks and keys date back to antiquity, the first being no more than a movable latch mechanism designed merely to keep the door *closed*. The earliest known example of a lock as we understand it today is the Egyptian door lock, so called because it was found among the ruins of the palace of Khorasabad in Nineveh. The Egyptian door lock was a pretty simple device as things go. A wooden housing contained a bolt with wooden pegs of different sizes along its length. When a long wooden key with another set of pegs was slipped into the bolt, the corresponding pegs on the bolt lined up, permitting the door to be opened.

The Romans used Egyptian principles when designing the first metal lock, then went on to invent the ward lock. (Wards are projections around the keyhole that prevent the lock from being turned without the proper key.) This kind of lock remained relatively easy to pick until Europeans started making their wards more complex. A case in point was Robert Barron's double-tumbler lock, invented in 1778, which relied on two tumblers of varying heights—a major advance in its time. But locksmithing didn't truly come of age until the Englishman Joseph Bramah put his hand to it.

In 1784, Bramah posted a sign outside his Piccadilly shop announcing a new lock of his own design, along with an offer of two hundred guineas to anyone who could pick it. Bramah was an accomplished inventor who had done his homework, and the reward went unclaimed until 1851, when an American visitor to the London Crystal Palace Exhibition finally conquered the device. The visitor who did this, and in the process made locks an object of pop fascination, was Alfred C. Hobbs.

Born in 1812 in Boston, Hobbs was only three years old when his father died and, as a result, he spent his early life in a succession of odd jobs. By the age of sixteen, he had already been a farmer, a dry goods salesman, a wood carver, a carriage-body maker, a carriage painter, a sailor, a tinker, a harness maker, and a glass cutter. All of this ranging around eventually began to pay off, however: as a glass cutter, he invented a method for securing a glass doorknob into its socket. Following on this success, he became partners in a Boston lock-making business called Jones & Hobbs, but he soon sold out his share to take a job with Edwards & Holman, which had just opened a lock-and-safe store in New York.

At Edwards & Holman, Hobbs finally received the education he needed. "During the time thus employed," he wrote, referring to himself in the third person, "the construction of locks was carefully studied and those locks that others valued seemed worth but little to him." Knowing that a bank would never buy a new lock unless its current one was shown to be defective, he designed a set of fine tools for the picking of locks and went off to close a few sales.

His first major victory came in 1847, in Stamford, Connecticut, at a bank that guarded its valuables with a Jones padlock, a ward lock, and an iron strap bolted over the keyhole. The bank directors agreed that if Hobbs could open their safe within two hours, they would pay him $150. He opened it in twenty-three minutes.

Hobbs became bolder as time went on. The following year, he was reading a newspaper in Lancaster, Pennsylvania, when he noticed a challenge from one Mr. Woodbridge of Perth Amboy, offering $500 to anyone who could open his lock within thirty days. At the moment, the lock was protecting a safe in the Merchants' Exchange reading room in New York.

"That's my money," Hobbs told a bystander, and promptly booked a train ticket.

On arriving in Manhattan, Hobbs discovered that Woodbridge had borrowed money from his father to finance the stunt. Not wanting to take money that didn't really belong to Woodbridge, Hobbs encouraged the man to surrender the certificate, which was waiting inside the safe, and to call off the stunt then and there. Woodbridge laughed him off and said he would risk it.

As well he could afford to: Woodbridge's lock was constructed in such a way that if the bolt was withdrawn before the tumblers had been completely aligned, anything lodged in the keyhole would be frozen in place, bringing the entire attempt to a halt. Hobbs, however, was aware of this construction and made new instruments accordingly.

The room was cleared at nine o'clock the next morning, and Hobbs set to work. Two and a half hours later, he had marked out all of the tumblers and was prepared to withdraw the bolt by pulling on a wire that hung from the keyhole. But that would have been too easy. Rather than announcing victory, he left the reading room and called on Woodbridge in the morning, with the news that something was wrong with the lock. Woodbridge smiled inwardly and agreed to meet the lock picker.

By ten o'clock, the room was filled with spectators. When the arbitrators made their appearance, Hobbs said nothing. Finally, Woodbridge arrived. Unable to make his way through the crowd, he called out to Hobbs, "What is the trouble?"

"There is something the matter with the lock," replied Hobbs.

"What is it?" asked Woodbridge.

Hobbs pulled on the wire, opening the safe. "Your lock won't keep the door shut."

From that point on, Hobbs was not only $500 richer but destined for a life of celebrity. He continued to sell bank locks, and as he began to travel more widely, he discovered that his reputation often preceded him. He also discovered that his profession created its problems, among them that his instruments could easily be confused with those of a burglar (Hobbs himself referred to them as "suspicious implements"). This disadvantage could be turned to good use, however. When Hobbs was planning a trip to the Crystal Palace Exhibition in 1851, he asked the London chief of police to send him a letter of introduction, not only to speed his way through customs but also to start the rumor mill turning.

Among the exhibits at the Crystal Palace was the Bramah lock, still unpicked after sixty-seven years. Hobbs had often heard of this lock during his travels, until gradually in his mind it had attained the aspect of a Holy Grail. As he packed up his exhibit of Day and Newell locks and sailed for England, he knew well that the challenge of his life awaited him.

The Day and Newell exhibit attracted considerable attention—too much attention, in fact. If the lock makers who pestered Hobbs with their endless technical questions weren't distracting enough, there was the Duke of Wellington, who kept bringing his friends—the queen, Prince Albert, and the Prince of Wales—to see the exhibit. Hobbs did his best to give his guests the slip and spent every spare moment in secret, studying the inventory of Bramah locks. But he was simply too well known, and eventually he was challenged to open a lever-tumbler lock, invented in 1818 by Jeremiah Chubb. Hobbs dispatched the Chubb lock in twenty-five minutes, to the amazement of his challenger.

Now there was no turning back. Hobbs stopped by the Bramah exhibit and, without revealing his identity, convinced a worker to let him have a look at the legendary lock. When the worker turned to serve another customer, Hobbs quickly produced a penknife and felt out the points inside the device. The worker caught sight of him doing this and asked what he was doing. Hobbs quickly explained that he was feeling to see if any of the parts moved. (He had heard that they did not—that the lock was a trap—but he now knew this to be untrue.) Tensions rose, and before long one of the proprietors arrived. Hobbs announced his intention of picking the lock but was put off until the following day, when he could meet with Edward Bramah (Joseph having passed on) and accept the challenge in person.

Together, the two men decided on terms that were as meticulous in their way as the cryptological efforts applied to computers today. The lock would be enclosed in a block of wood and screwed to a door, the screws sealed and the keyhole covered with an iron band, so that no one but Hobbs could have access to it. If Hobbs succeeded in picking the lock, the key would be used to lock and unlock the padlock again, as proof that no force had been used. Three arbitrators would see to it that these conditions were met.

It was July 24, 1851, when Hobbs first went at the Bramah lock. Discounting time when he was taken away by other business, he sweated for fifty-one hours, accumulated over the course of sixteen days. On August 23, he exhibited the opened lock to the arbitrators. Six days later, he locked and unlocked it again, and on the following day the key worked perfectly. The Bramah lock, touted for years as being as "impregnable as Gibraltar," was impregnable no more.

Hobbs received his promised two hundred guineas, and his own shard of immortality in the bargain. Indeed, he became so popular that he could barely afford to turn down a challenge, for fear of being advertised as a coward.

The picking of the Bramah lock had its effects beyond the moment as well. For one thing, Hobbs helped boost the reputation of American innovators among Europeans, who before the Crystal Palace opened had held the prospects of the Yankee exhibits to be poor. The lock industry was also sufficiently energized in 1851 for Linus Yale Jr. to design the first Yale lock, which continues to set the standard today.

But perhaps the most unexpected turn came in show business. It was only a few short years after the Crystal Palace Exhibition, after all, that escapology became the rage on both continents, and but a few short years again before Harry Houdini began collecting what eventually amounted to a vast collection of keys and devices for the picking of locks for his fabulous stunts. This connection is surely worth noting because, for all the skills the twentieth-century hacker may boast, it was Houdini who possessed the truly important secret: *how to get out of jail.*

4
THE OFFICE

*The office, more than any other living space,
is supposed to be an efficient machine.
Yet even here, many products mask
a tale of woeful disarray . . .*

No. 782,181.   PATENTED FEB. 7, 1905.

G. O. SQUIER.
WIRELESS TELEGRAPHY.
APPLICATION FILED NOV. 10, 1904.

2 SHEETS—SHEET 2.

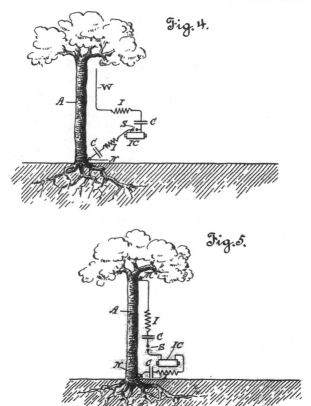

Fig. 4.

Fig. 5.

Muzak

Georwas born in Dryden, Michigan, on March his "wire-wireless" technology—which allowed radio signals to travel through a telephone line—would usher in the age of the "radiopolis." But after a series of losing battles against AT&T, it ended up as that emotion regulator known as Muzak instead.

Squier was born in Dryden, Michigan, on March 21, 1865, only days before the Civil War ended. Whatever his childhood may have been like, it does not figure highly in his memoirs. He was accepted into West Point in 1883 and, upon graduation, enrolled in Johns Hopkins University, with the aim of applying the latest scientific advances to the betterment of the U.S. Army Signal Corps.

An imaginative man in the strict culture of the military, Squier spent the first two decades of his career being shuttled from one project to another. The 1890s saw him magnetizing razor blades, investigating searchlights, and inventing the Squier synchrograph—a device that tracked the flight path of projectiles. While attending the 1893 Columbian Exposition in Chicago, he met the scientist William H. Preece, who, a few years later, was able to introduce him to the young Guglielmo Marconi.

The incident must have left an impression, because by 1897 Squier had designed his own radio and was using it to ring bells, fire cannons, light lamps, and detonate mines by remote control. Concerned that radio equipment was too cumbersome on the battlefield, he next set his mind to making radio antennae from "living vegetable organisms." The trick turned out to be simple enough: two spikes, one driven into the base of a tree, the other driven in further up, tapped into the natural conductivity of the trunk. On February 7, 1905, Squier received a patent for his "floroscope"—and, one imagines, earned the exasperated sighs of his superiors.

Squier was a tireless supporter of other inventors as well. In September 1908, when Orville and Wilbur Wright demonstrated their airplane for the army at Fort Myer—the first demonstration for the American public—Squier was there to greet them. He got on one of these flights himself, which made him the second airplane passenger in history.

Soon after the Fort Myer flights, Squier received an assignment to investigate the state of army radio technology. In those early years, radios worked by a spark gap: an electric charge jumped through the air, completing a circuit and discharging an electromagnetic wave. Unfortunately, these sparks had caused interference in nearby radio equipment. Another drawback was their low power output. The Signal Corps needed radio sets that could transmit up to one hundred miles, and spark transmitters could scarcely manage twenty. But Squier, who had worked with many different kinds of technology, thought he saw a way out. What if it were possible, he wondered, to run radio signals over existing telephone lines? Wouldn't that keep the signal insulated from other signals around it and, in the process, give it greater range?

In 1909, Squier set to work in the Signal Corps laboratory at 1710 Pennsylvania Avenue. Soon he expanded into a new research laboratory in the Bureau of Standards, some seven miles away. For his phone line, he leased one belonging to the Bell Company that ran between the two labs.

On September 29, 1910, Squier was able to show his superiors the results. Not only did his team successfully send radio signals through the telephone wires; by using tuning circuits, they also sent simultaneous messages over the same telephone line. With this demonstration—that a radio signal could be changed into a telephone signal or a telephone changed into a radio signal, and that several conversations could travel through the same telephone line—a whole slew of problems had been solved at once.

The communications industry was quick to recognize the achievement. A committee appointed by the Franklin Institute declared wire-wireless "a distinct contribution to the field." *Telephony, Telegraphy, and Wireless* magazine called it the "greatest advance made in electrical communication since the introduction of the telephone itself." John Stone Stone, an accomplished (and doubly solid) communications engineer, proclaimed that "a new art has been born to us. . . . It is certainly a most promising youngster and should, after the manner of its kind, call lustily for its share of attention and sustenance."

But then Squier delivered the real shocker: rather than convert his invention into a fortune, he had chosen to give his patents to the public.

Big mistake. With the coming of World War I, Squier was assigned to other more pressing matters. AT&T, meanwhile, used the time to improve wire-wireless, as his invention was then called. It bought up the rights (using some deception) to Lee De Forest's cutting-edge vacuum tube. It hired George Campbell to design a filter that eliminated what interference was left on the phone line. And it happily used the invention that Squier had given away. By the time the Armistice was signed, AT&T had nearly perfected wire-wireless for its own use.

Squier was amazed at the phone company's results but disturbed that the "public" might turn out to be one corporation and no one else. Worse, AT&T executives were now openly doubting Squier's priority in wire-wireless. They had some patents of their own, they said, that predated his. After careful consideration, Squier decided to sue the phone company for infringement. The case turned on a technicality: Squier had offered his invention to "any other person in the United States." The question was, Did that mean any other American citizen, or any other person in the United States government?

As the case against AT&T wended its way through the courts, the two parties went on their respective mad dashes to make wire-wireless a phenomenon. Squier, for his part, designed a radio set that plugged directly into the wall and received its signals through electric power lines. In 1922, he approached the North American Company, a

Cleveland, Ohio, enterprise, and arranged to have the initial tests for his system conducted in the aptly named Ampere, New Jersey.

The following year, Squier retired from military service and looked on with satisfaction as North American installed a radio-transmitting service over the electric lighting lines in Milwaukee, Wisconsin. The principle resembled that of modern-day cable television: a central station received and rebroadcast a variety of programs, with a bill of fare that included news and dance music. Muzak, then known as Wired Radio, made its quiet appearance in the world.

AT&T went in another direction. Until the advent of satellite and advanced submarine cable technologies, long-distance radio signals had to be sent by telephone lines from a central station to stations in hub cities, where the signal would continue on as a radio wave to listeners' homes. Thus AT&T was in a central position to determine the fate of radio. It could lease its phone lines to radio stations around the country, exactly as it leased its lines to telephone subscribers. In a stroke its competitors became its debtors.

On September 4, 1924, the court announced that it had found against Squier. Still struggling to get Wired Radio off the ground, Squier immediately appealed the decision, but it was to no avail. "Pride in the army and his own corps and a very human liking for the limelight," remarked the appellate judge sitting on the case, "all produced in Maj. Squier a desire, quite sincere at the time, to do exactly what the subtitle of his patent indicated and have it 'dedicated to the public' . . . of course, the very existence of this suit shows that the fit of public spirit has passed." In the end, Squier's only solace was the court decision to shunt a portion of AT&T's profits to the National Academy of Sciences.

Still, Squier was not ready to bow out of the game. In 1926, he patented the Monophone, a variant of the wire-wireless designed to send radio programs over telephone lines to rural listeners. By 1930, Wired Radio was broadcasting three channels to listeners in the old Lakeland district of Cleveland, Ohio, at a cost of $1.50 a month. And on Squier's recommendation, programmed music began to flow into the dining rooms of hotels and restaurants.

Wired Radio was eventually forced to abandon its electric lighting lines, which suffered from excessive interference, in favor of the only other option: leasing telephone lines from AT&T. The company survived, but not with Squier at the lead. It

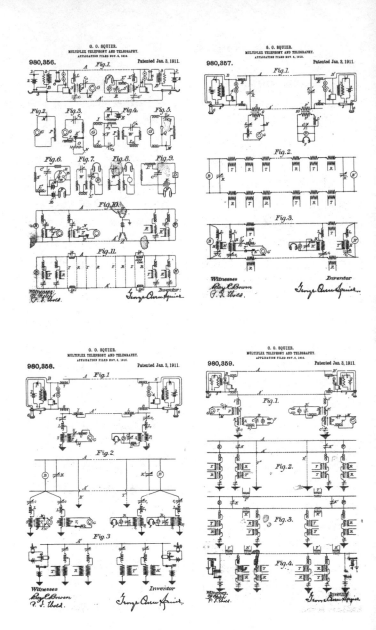

was his successor—Waddill Catchings, a Wall Street investment banker and early promoter of consumerism—who saw the potential of wired radio systems in department stores and dentists' offices and, in the 1930s, began to develop programming that increased worker productivity even as it dulled the mind.

Squier had been acquiring land around his grandfather's farm in Michigan since the turn of the century. By the 1930s, these holdings amounted to a great deal. With his career on the wane, he dubbed his estate "The Poor Man's Country Club" and invited anyone and everyone to come and hunt, fish, or play golf on his grounds. In the end, more than sixty thousand guests visited annually. It was in this context that Squier waxed expansive in a book called *Telling the World,* which predicted a future filled with "radiovas" (radio opera singers), radio streets, radio cities, radio everything. It seems that Squier had been studying his futurists.

By this time, of course, wire-wireless was paying off in spades for AT&T. In 1927, it inaugurated the first transatlantic telephone service to London using two-way radio. A similar service to Hawaii began in 1931, with Tokyo following in 1934. In 1937, AT&T developed a superhigh-frequency wire capable of delivering high-definition television images, and radio stations eager to get into commercial television happily signed to lease its telephone lines. Within a few years, the television age was in progress.

George Owen Squier died of pneumonia on March 24, 1934, and was buried with full military honors in Arlington Cemetery. Today, the air above his grave—indeed, the air everywhere—carries an untold chorus of voices to and from telephone lines. As for Muzak, one could almost chart the course of the American century by the spaces it has saturated. It wafted through the deserted halls of the United States Embassy after the last Americans had left Saigon. It played in the cabin of Apollo 13 during its benighted lunar mission. And in the end, it found its most intractable use exactly where it began: in the pumping of radio signals through the telephone line . . . every time you're put on hold.

The Pencil

In 1565, Konrad Gesner was a medical scholar living in Zurich. Gesner had already written a treatise on the virtues of milk and was moving on to his next attempt: a bibliography of all the recorded knowledge in the world. In the midst of this monumental task, he came across a new kind of writing device, which appeared to be a cylinder of lead sheathed in a wooden case. Who exactly the inventor of the pencil was, Gesner did not say. He simply made a note of it—with his quill, the current instrument of preference—and continued on to the subject of fossils. The following year, he died of the plague.

There is something perfect about this story, with its pairing of small and large. A man sets out to embrace the whole of the human experience and never gets anywhere close, yet the modest object that falls under his gaze goes on to become . . . well, instrumental to the recording of knowledge from that moment on. Explore this loop a little further and the implications turn downright screwy: Gesner was writing about a method of writing, and for as much as he thought he was writing the ultimate document, today a new slew of documentarians (the present author included) are taking their turn in writing about him. The pencil, it seems, is continually creating more work for itself.

That said, the pencil does have a history that moves forward in time and gets somewhere. Its origins lie in antiquity, when Egyptians used a small lead disc for making guidelines on papyrus. (The actual writing was done in ink.) The Greeks later picked up on this practice, as did the Romans, who called the disc a *plumbum*—Latin for *lead*.

Not much happened in the way of pencil technology for another millennium after that. Then, in 1564 (the year before Gesner mentioned the pencil), a deposit of graphite was discovered at Borrowdale in Cumbria. This graphite—called plumbago because it acted like lead—was so solid and uniform that it could be sawed into sheets and then cut into thin square sticks. As the only pure graphite deposit ever found, it also held fantastic financial promise. Less pure deposits of graphite were available in many parts of the world, but they had to be crushed and the impurities removed. And by then the graphite tended to crumble.

The Borrowdale mines were active for only six weeks every year, and after the wagons were filled with the stuff, armed guards escorted them to London. Export of the ore itself was prohibited. Instead, it was routed to a newly formed guild—the English Guild of Pencilmakers—which carved wooden cases for the graphite sticks and, in a separate development, enjoyed a world monopoly on the sale of the finished product. Indeed, one could say the English Guild of Pencilmakers was the Microsoft of its day. One could also say that there is a direct line of descent between them: the monopolies of Elizabethan England, most notably the textile monopoly of the Merchant Adventurers, eventually evolved into the joint-stock companies, such as the Virginia Company, that originally footed the bill for the colonization of the New World.

Of course, give people pencils and eventually they'll start to have ideas, no matter how strong your monopoly is. By the seventeenth century, the Germans were using a mixture of graphite, sulphur, and antimony, and the resulting white lead sticks were said to compare favorably with the English pencil. In 1779, K. W. Scheele made a chemical analysis of plumbago that proved it to be a form of carbon, not of lead. A decade later, A. G. Werner suggested the more appropriate name *graphite*, from the Greek word meaning *to write*. But as is often the case, it took the privations of war to bring about the decisive change.

In 1795, when France was cut off from both the English and German pencil sources, Napoleon commissioned Nicolas-Jacques Conté, an officer in the French army, to develop a viable substitute. Conté mixed powdered graphite with clay, then fired the mixture in a kiln. This method was not only serviceable, but allowed the sticks to be graded from hard to soft by varying the proportion of graphite to clay. When the Napoleonic Wars ended, this new method spread abroad and was eventually adopted by all pencil manufacturers.

In an interesting postscript, the isolation of Napoleon's troops also led to one of the first significant uses of the new pencil. After marching on Egypt, Napoleon decided to withdraw most of his troops but left behind a draftsman named Dominique-Vivant Denon, who proceeded to document the various wonders of that land. This could be bracing work: in the midst of one of his sketches, Denon paused just long enough to reach for his pistol and shoot an attacking Arab. (When it was later pointed out that the dimensions of this sketch were incorrect, Denon blamed the attacker for knocking his easel askew.)

Denon finally returned to Paris and showed his sketches to Napoleon, who liked them so much he appointed Denon as director of the Louvre, which, not coincidentally, soon housed many of the treasures that Napoleon had looted from Egypt. In time, the Louvre became a favorite location for art students, who faithfully wore down their own pencils sketching these very same treasures. Thus an Egyptian invention found its way, if not home, at least into familiar surroundings.

The Electrical Outlet

T he story of the electrical outlet begins in the spring of 1881, when a twenty-five-year-old Serbian named Nikola Tesla failed to show up for work at the Central Telegraph office in Budapest. Tesla was certainly ill. He claimed to be able to hear a housefly land on a table and a ticking watch from three rooms away. His pulse boomeranged from 260 beats per minute to a sluggish pace. In the dark, an object twelve feet away gave him "a peculiar creepy sensation on the forehead."

After his health returned, Tesla began taking constitutionals in the city park with his friend Anital Szigety. One evening, as the sun set, Tesla was reciting some of his favorite lines from Goethe's *Faust* when he stopped abruptly and stood dumbfounded.

Szigety thought his friend might be having a relapse and urged him to sit on a bench to recover. But Tesla would not rest until he had found a stick and drawn a diagram in the dirt. "See my motor here," he said breathlessly. "Watch me reverse it."

What Tesla sketched out was a diagram for an alternating-current electrical motor. At the time, electricity was still an unharnessed power. Thomas Edison, having recently invented the lightbulb, was busy designing an electrical system based on direct current—the type of current used in batteries today—but direct current could not travel very far before it dissipated. Alternating current would be different. It could travel along wires for great distances without appreciable weakening. With a few strokes in the dirt, Tesla had discovered a way to make AC electricity feasible. His troubles, however, were only just beginning.

Tesla went to Paris in 1882 and tried to sell his idea to the Continental Edison Company, which was then headed by Charles Batchelor. Though Batchelor turned the invention down, he saw promise in Tesla and hired him as a regular employee. On his own time Tesla built his first AC induction motor and became a professional billiards player to make ends meet. Eventually, though, Tesla prevailed on Batchelor to write a letter to Edison, recommending his transfer to New York. And so in 1884, at the age of twenty-eight, Tesla stepped off a boat with nothing in his pocket but his notes for the AC motor, a few poems and articles, a diagram for a flying machine—and a letter of introduction to Edison.

Proceeding directly to Edison's offices at 65 Fifth Avenue, Tesla introduced himself briskly, then launched directly into his alternating-current scheme. Edison scoffed at the idea, having staked his claim on DC, and put him to work that day on a lighting plant. A frightfully diligent man, Tesla worked from ten-thirty in the morning to five the following morning every day. Soon he found a way to improve Edison's existing DC systems, and he reported as much to his boss. Edison allegedly answered, "There's fifty thousand dollars in it for you—if you can do it." When Tesla made the improvements, however, Edison laughed in his face, saying, "You don't understand our American sense of humor." Edison's side remembered a different story with the same ending: Tesla had offered to sell his AC patents for $50,000, and Edison had turned him down with a laugh.

Whichever story is the correct one, Edison would live to regret the loss. In April 1887, Tesla was able to set up his own company at 33–35 South Fifth Avenue (now West Broadway), not far from Edison's headquarters. The following year, he received his first AC patents, and on May 16, 1888, he was invited to lecture at the American Institute of Electrical Engineers.

One member of the audience that day was George Westinghouse, a portly tycoon of eminence who had become rich from his own patent for railroad air brakes. On July 7, 1888, Westinghouse paid a visit to Tesla's Greenwich Village lab, found the machinery to his satisfaction, and struck a deal. Tesla would receive $2,000 a month to act as a consultant and to work on Westinghouse's burgeoning AC system.

But Tesla's fate was not yet assured. Another man who witnessed the AIEE lecture was Harold Brown. A staunch opponent of AC electricity, Brown soon appeared at the gates of Edison's lab in West Orange, New Jersey, where he discovered the Wizard already hard at work on a negative campaign against AC technology.

The field general in this crusade, which was meant to publicize the dangers of AC and which came to be known as the War of the Currents, was Samuel Insull. Perhaps as early as 1887, Insull had begun paying neighborhood boys a quarter to round up stray dogs and cats. Then, in the presence of newspaper reporters and other invited guests, these unclaimed pets were lured onto a tin sheet and subjected to 1,000 volts of alternating current.

Various names were floated for this type of execution: ampermort, dynamort, electricide. After the Tesla-Westinghouse agreement was struck in June 1888, the press settled on the verb "to be Westinghoused."

On July 30, 1888, Brown gave his own demonstration at the School of Mines at Columbia University, before an audience of electricians, reporters, the superintendent of the SPCA, and a handful of public observers. Falsely declaring himself a disinterested party, Brown compared and contrasted the effects of AC and DC on a large Newfoundland mix that had been attached to a pair of cables. First, he administered DC in increasing doses. The caged dog's whimpers turned to yelps, then to expressions of surprise, anger, fear, and agony. At 1,000 volts, it had not yet died. Some members of the audience had already left the room by the time Brown promised to make the dog "feel better." With a quick jolt of no more than 330 volts of alternating current, he killed it.

When administering the DC current, he made use of a relay that shut the juice off as soon as it was delivered; the AC current, however, had no such relay. Though no one in the audience discovered the trick, one skeptic suggested that the DC doses had simply worn the dog out. Showman that he was,

Brown called down to the basement for another victim. At this point, the superintendent of the SPCA stepped in and prevented a second execution.

Meanwhile, the New York State legislature had been considering proposals for a humane method of capital punishment and, out of some thirty-three choices, had settled on electrocution as the way of the future. The Medico-Legal Society of New York chose Brown and his assistant at the Columbia gala, Dr. Peterson, to perfect a method. Under the watchful eye of Edison, the two men electrocuted and dissected about two dozen dogs, two calves, and a horse. According to the *New York Times*, the calves' meat was pronounced "fit for food." The society's report, delivered on December 12, 1888, came out in favor of the electric chair.

At this point, Edison joined with Elihu Thomson—an inventor who had also staked his claim on DC electricity—in a masterpiece of industrial espionage. Brown, now an electrocution "expert" for New York State, wrote to Edison, asking him to foot the bill for the generators that would electrocute a human being at Sing Sing penitentiary. As Westinghouse would clearly not be quick to sell the necessary equipment to his adversaries, Edison worked with Brown through the Thomson-Houston Electric Company, which worked through a third party to procure a few Westinghouse generators that were being sold to make way for upgraded models. Cryptically, Brown boasted that the generator he purchased "already had a record as a man-killer."

The date was set for August 6, 1890, and the privilege of being the first human to die in an electric chair fell to a convicted murderer named William Kemmler. People tried every form of persuasion to get one of the twenty-five witness seats. The Whitechapel Club of Chicago expressed a desire to buy Kemmler's body when the ordeal was over.

Before the execution, the witnesses were locked into the chamber. Kemmler entered dressed for a day at the track, wearing a gray jacket, a checkered shirt, a bow tie, yellow pants, and polished shoes. After he was seated and strapped in, the executioners discussed how long the voltage should be applied. Hastily deciding on ten seconds, they threw the switch. A Dr. MacDonald started the stopwatch, and Kemmler's body went taut against the straps.

Fifteen seconds later, MacDonald cried, "Stop!"

The prisoner had no pulse. But when the doctors began to remove him from the chair, someone noticed that a cut was still sending out a trickle of blood. Kemmler's heart was still beating.

The audience let out a mixture of groans and

sobs. The prisoner was strapped in again and subjected to a considerably longer dose, causing bloody sweats, gurgles, a heaving chest, burning skin, and the smell of feces. Afterward, the body was hot to the touch. One of the physicians offered the simile "like overdone beef."

Edison never admitted to any direct involvement in the execution. "I have not failed to seek practical demonstration," he insisted. "I have taken life—not human life—in the belief that the end justified the means." But maybe he was being so defensive because he himself had come under attack—not physically but economically.

In 1888, while the three major electrical leaders of the time—Elihu Thomson, Westinghouse, and Edison—waged the War of the Currents, J. P. Morgan had begun working behind the scenes, trying to break each of them in turn and bring them under his dominion. By initiating price wars, Morgan weakened the position of the Thomson-Houston Electric Company, then put Charles A. Coffin at its helm. Coffin then proceeded to Westinghouse's office, where he met with less luck. Listening to Coffin brag about what he had done to Thomson, Westinghouse replied, "You tell me how you treated Thomson and Houston. Why should I trust you?"

Edison was no more willing to sell out than Westinghouse was. When the idea of a merger with

Thomson-Houston was broached, Edison wrote to the president of Edison General Electric, Henry Villard. "I can only invent under powerful incentives. No competition means no invention."

Unfortunately, Villard, acting against Edison's wishes, secretly signed Edison General Electric over to Drexel-Morgan, and on April 15, 1892, the company merged with the Thomson-Houston Electric Company to form a new entity called General Electric. When Edison learned that even the Edison Lamp Company had fallen under Morgan's sway, he turned pale and sent out a statement that he had been "gulled."

In the end, only Westinghouse had come out of the battle intact. Indeed, by the time of the Columbian Exposition of 1893, the War of the Currents was essentially over, and alternating current had become the clear winner. In the Electricity Building, Edison used the technology to power all three thousand incandescent bulbs in his Pillar of Light. General Electric, by then Elihu Thomson's employer, supplied the three-ton searchlight that ran on alternating current. Even Westinghouse had entered the fold and licensed his AC patents to his sworn enemy, Samuel Insull, to provide the juice for the grounds, the fountains, and the waterways.

But there was still one point left to be made. As

one witness who stumbled onto a modest space in the Electricity Building described it:

> Mr. Tesla has been seen receiving through his hands currents at a potential of more than 200,000 volts, vibrating a million times per second, and manifesting themselves in dazzling streams of light. . . . After such a striking test, which, by the way, no one has displayed a hurried inclination to repeat, Mr. Tesla's body and clothing have continued for some time to emit fine glimmers or halos of splintered light.

Clearly, if Tesla could illuminate himself with 200,000 volts of alternating current and live, then Harold Brown's claims of its dangers were baseless!

In October 1893, just after the Columbian Exposition closed, the Niagara Falls Commission chose alternating current to power its new electrical plant. Westinghouse would fire its new AC generator with 15,000 units of horsepower. General Electric, meanwhile, would provide the distribution lines. Alternating current began to course through a dense web of high-tension wires, all the way from who knows where, and into those little rectangular structures that began to appear on the walls of offices and homes.

5
THE GARAGE

Theoretically, every invention in this book could find its way into the average garage. This is the graveyard of commerce, where all good products go to decompose . . .

3,902,106

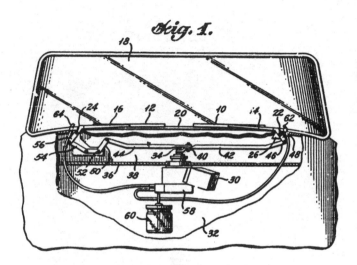

Fig. 1.

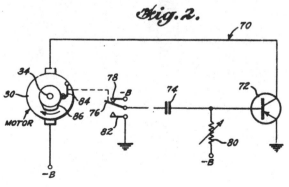

Fig. 2.

MOTOR

The Intermittent Windshield Wiper

Robert Kearns was celebrating his wedding in 1953 when he opened a champagne bottle and the cork flew into his left eye, causing permanent damage. Another person might have taken this accident as a cautionary tale, but Kearns was capable of adding imagination to injury. Nine years later, he was driving in a light rain, squinting at the windshield and thinking about his bad eye. Windshield wipers in those days came in only two speeds, fast and slow, and unless it was raining reasonably hard neither one was much help. Musing on these two unrelated facts, Kearns began to wonder whether a windshield wiper could be designed to operate like an eyelid—leaping into action only when necessary.

Kearns had tried his hand at inventing before. His first effort had been a comb that dispensed its own hair tonic. When that didn't get past the model stage, he went through a scattering of ideas—an amplifier for people who had undergone laryngectomies, a new kind of weather balloon, a navigational system—none of which amounted to much. In fact, his success rate was virtually nonexistent, but for an inventor, failure is simply preamble to the one great idea.

Kearns started designing his windshield wiper in the spring of 1963 and had a working model before the summer was out. As he had conceived it, his wiper could remain at rest for varying lengths of time, sweep across the windshield at varying speeds, and even respond to the amount of moisture present. A native of Detroit, Kearns had long stood in awe of the Ford empire, so it was only natural that he take his invention there first.

The Ford engineers were impressed when Kearns demonstrated his prototype in his own Ford Galaxie. All he had to do, they said, was run a few tests to prove that his wipers met industry standards. On November 16, 1964, Kearns was able to report success. By this time, however, the enthusiasm at Ford had waned. Desperately broke, Kearns approached David Tann, co-owner of a small tool-and-die shop. Tann loved Kearns's wiper and agreed to prosecute the patent. In the meantime, he said, he would pay Kearns $1,200 a month to continue his research.

After Kearns received a patent for his wiper, Ford turned around and offered him a job on its engineering team. That was about as good as it got. Five months later, he was laid off with little explanation. Worse still, his intermittent windshield wipers began to appear on Ford cars, despite the care he had taken to retain his patent rights. General Motors followed suit after that, then Chrysler, Saab, Volvo, Honda, and Rolls-Royce. One day, Kearns took apart a windshield wiper made by Mercedes and discovered his own invention inside. Almost the entire automotive industry had stolen his idea! According to his son, Kearns's hair turned completely white that day.

Another type of person might have given up right there, or perhaps applied pressure until the infringers agreed to settle out of court. Kearns, for his part, vowed to fight it out to the bitter end—and through the bitter middle as well. A man with no legal background, he decided to represent himself in court and to enlist the help of his family in preparing his case, which ended up lasting thirty years. As his daughter Kathy put it, "The lawsuit is all we've ever known."

The atmosphere was not exactly easy on the

Kearns family. At one point, Robert realized how badly he'd been neglecting his children, bought two kites, set off on foot from his home in Gaithersburg, Maryland, and promptly wandered off into parts unknown. He was found by police in a park in Tennessee, the kites still in his hands. After years of such behavior, Robert's wife, Phyllis, threw up her hands in despair and left him.

Even when victory was within reach, Kearns almost let his obsessive nature get the best of him. In January 1990, Ford agreed to settle for $30 million. On principle, Kearns turned it down. A second trial commenced, during which Kearns thought it appropriate to disappear to the Little Bennett Regional Park in West Virginia, where he subsisted for a time on a diet of pork and beans. Only when the judge threatened to initiate proceedings on the plantiff's mental competency did Kearns rouse himself and come in from the cold.

The day of reckoning did finally arrive. Kearns received a settlement from Ford of $10.2 million. In 1992, in a separate case against Chrysler, he fetched another $11.5 million. As the gavel came down, the judge looked over to Kearns and remarked, "I'm sure I will see you here again."

And so he probably will. Rather than rejoicing in his victory, Kearns expressed his dismay that the jury had found no willful infringement on the part of Chrysler. One can hardly blame him. In the annals of technology, tales of corporate abuse are strewn across the American century like so many highway accidents, and in a sense, Kearns has fought for the many inventors who ended up as roadkill. Then again, a psychologist might reach for a different metaphor to describe the story of the intermittent windshield wiper. Maybe what Kearns is looking for is not vindication so much as the equivalent of the event that started it all: a quick champagne cork in the eye.

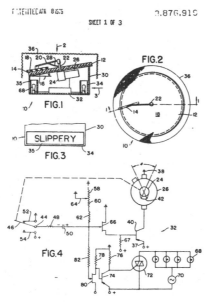

The Standard Screw Thread

T
he standard screw thread is so much a part of our lives as to be almost beneath comment. Boring even, you might say. It didn't start out that way, though. The modern screw began with the dream of perfectly lifelike robots.

To be more exact, it began with a Frenchman named Jacques de Vaucanson. Born in 1709 in Grenoble, France, Vaucanson was the youngest of ten children and, like so many runts, discovered that he would have to work hard if he wanted to get any attention. He worked hard. In 1725, he entered the Order of Minims in Lyons, but his idea of heaven turned out to be more mechanical than mystical. When the priests broke into his make-shift workshop, they discovered that he was developing a series of flying angels, which they promptly destroyed.

Vaucanson returned the favor by leaving the order and becoming the greatest inventor of automata who ever lived. Moving to Paris, he quickly made up for lost time in the brothels, which may account for his burgeoning interest at that time in both music and anatomy. In any event, he was soon attempting to construct what he called "moving anatomies." This project put him deeply in debt, but it also laid the groundwork for his masterpieces, which—not coincidentally, perhaps—he made strictly for financial gain.

Vaucanson's first automata, begun in 1735, resembled the suitably romantic figure of a flutist. A life-sized figure dressed in rustic garb, this machine blew air and moved its fingers in such a way as to produce eleven different melodies on a flute, and when it was unveiled in October 1737 at a fair in Saint-Germain, all of Paris flocked to see it. In the same year, Vaucanson completed two other automata. One was a "shepherd" and a bit of one-man band; it played from its repertoire of twenty melodies on the flute and drum both. But it was the third android that topped them all: "an artificial duck of gilt brass which drinks, eats, flounders in water, digests and excretes like a live duck." Capitalizing on his knowledge of anatomy, Vaucanson constructed his duck with openings that allowed the public to look in and observe its gastric orchestrations. What became of the food at the other end is generally not mentioned.

So much for the old maxim about what something must be if it looks like it, and so on. In fact, with the passing of time, Vaucanson's masterpiece (which, oddly, could not quack) started to be almost everything but a duck. For one thing, his attention to precise anatomy led him to create the first tubes made of rubber (known at the time as *caoutchouc*). In slightly modified form, the same invention can be seen under the hood of any car today.

Vaucanson didn't really begin branching out, though, until 1743, when—wealthy, famous, and restless—he sold all three automata. By this time, he had been appointed to the French silk factories, and, after observing inefficiencies there, he constructed a device for weaving brocades automatically. This loom, the invention of which is often attributed to Jacquard, provided the inspiration for Charles Babbage a century later when he was devising the first mechanical computers.

But to return to the screw. Vaucanson continued to be prolific throughout his life, and his later inventions invariably built on the principles of the earlier ones. One example of "continuity-think" was his industrial cutting lathe with prismatic guideways, invented in 1760. In constructing his automata, Vaucanson had learned the virtues of extremely precise tooling, and this lathe allowed machinists to cut more precisely than ever before. As it turned out, it also helped an Englishman named Henry Maudslay along in his career.

Maudslay was not at all the wild man that Vaucanson was. Born in Woolwich in 1771, he grew up around the dockyards, where his father kept a store, and he began work as a "powder monkey" at an early age, making and filling cartridges for the local arsenal. At the green age of thirteen, he caught the eye of the famous locksmith and plumbing genius Joseph Bramah. But Maudslay was too bright to abide another genius very long. When Bramah refused to give him a raise, he struck out on his own.

By 1797, Maudslay had set up his own shop and developed a slide rest lathe, which improved on Vaucanson's lathe in both the speed and the precision with which it could cut metal. Of the slide rest lathe, James Nasmyth offered that "its influence in improving and extending the use of machinery has been as great as that produced by the improvement of the steam engine in respect to perfecting manufactures and extending commerce."

No small claim, coming from the inventor of the

steam hammer. Was it justified? Probably, considering that Maudslay's lathe, which incorporated a blade of crucible steel mounted on accurately planed triangular beams, allowed him to do work on a large scale while retaining the locksmith's or the clockmaker's precision. By 1808 Maudslay, together with the Frenchman Marc Isambard Brunel, had established the first real mass-production unit at Portsmouth and were turning out wooden rigging blocks used in hauling and, in particular, aboard naval ships for moving guns rapidly into position for firing. A vessel of the third class required 1,400 blocks to operate, all of which, up to this point, had been made by hand. But an order like that was no problem for Brunel and Maudslay, who could produce *130,000* of the things in a year.

Indeed, Maudslay's work opened the way for the making of interchangeable parts, a prerequisite for any sort of mass production, and he soon became famous. In 1810, he moved his works to Lambeth, where he went on to improve the original slide lathe and to invent new machine tools, including an improved micrometer. Many future engineers came to work for him as apprentices, among them a man named Joseph Whitworth.

Henry Maudslay died in 1831, but not before

Whitworth, working at his footstool, developed measuring instruments accurate to within a millionth of an inch.

A number of people had claimed success in creating interchangeable parts, among them cotton gin inventor Eli Whitney and the revolver man Samuel Colt. But no one had actually done it. Neither Whitney nor Colt ever progressed past the stage of idle boasts, and Vaucanson himself had only gotten so far before he lost patience. It took a yeoman like Whitworth to abjure glamour and buckle down, and in 1841 he was able to announce the realization of the standard screw thread in a paper titled "A Uniform System of Screw-threads." The first standardized screws soon followed.

There are a number of curious morals to be fished out of this tale. The famous scientist Hermann Helmholtz caught one of them when he looked back on the entire progression from his vantage point of the late nineteenth century and stood awestruck at the work of Vaucanson. The early French automata, Helmholtz believed, represented a major step forward in precision technology, even if they ultimately failed at re-creating life. But, he went on to note, "nowadays we no longer attempt to construct beings able to perform a thousand human actions, but rather machines able to execute a single action which will replace that of thousands of humans."

Of course, we've come a long way from the automation nightmares of Helmholtz's day. As a production machine, the personal computer represents a major departure from the Industrial Age, because it allows users the freedom to create their own style of working—perhaps enough freedom, say, to devise a generation of androids more lifelike than anything Vaucanson ever dreamed of. Then again, look at your computer the next time you have a chance and ask yourself, What holds it together?

The Flat-Bottomed Paper Bag

I
t has often been noted that inventions tend to grow out of wartime research. Less frequently noted is the fact that inventions will sometimes appear in the vacuum war leaves behind. This was certainly the case during the Civil War. When the men went off to do battle, the women had to make do with meager resources, and they responded by filing almost twenty times as many patents as they had the previous decade. Perhaps the most successful of this new breed of inventor was Margaret Knight, inventor of the flat-bottomed paper bag.

Knight showed an aptitude for the workshop early. While other girls were learning to cook and sew, she was learning to make sleds, kites, and whatever else her brothers requested. Her childhood was not without its difficulties. As she later put it in an 1872 interview in *Woman's Journal:* "I was called a tomboy; but that made little impression on me. I sighed sometimes, because I was not like other girls; but wisely concluded that I couldn't help it, and sought further consolation from my tools."

Sometime around 1867, after a decade of odd jobs, Knight took a job at the Columbia Paper Bag Company in Springfield, Massachusetts, at a third of the standard wage. The low pay was what you might call a fashion statement: the superintendent didn't like to hire women because, as he saw it, their hoop skirts got in the way of the machinery. Given that observation, Knight might have chosen to re-design the hoop skirt; instead, she decided to im-prove the machinery.

From all accounts, she did quite a job of it. She immediately began using her time at the Colum-bia factory to experiment with more efficient ways of making paper bags, causing one worker to com-ment that "she had a sharper eye than any man in the world." Before long, she was able to wow her fellow employees with a machine that cut, folded, and pasted bag bottoms by itself. When her em-ployer complained that she was spending too much time experimenting, she offered him a percentage of future royalties. That ended the complaints from her employer. Meanwhile, she progressed from her original wooden model to an iron one, which she intended to send to Washington for patenting. (Until 1880, the Patent Office required a model no larger than one cubic foot to accompany every patent application.)

It was at this point that Knight encountered her first serious obstacle. A visitor to the machinist's shop named Charles F. Annan had taken a keen interest in her accomplishments. Too keen, it seems: before she was able to send her model to the Patent Office, Annan was already manufacturing a similar machine. Knight, struggling to make ends meet with a moonlighting job in real estate, promptly brought an interference suit against Annan.

Of course, inventors have been going to court since there were courts, but this case was different, because it was the nineteenth century and one of the parties was female. Annan's lawyers capitalized on this second fact, insisting that Knight, frail thing that she was, could not possibly have achieved what she claimed.

Another woman might have crumbled under the assault, but Knight was prepared to give as good as she got. She produced 1,867 drawings and explained them in such detail that no one could doubt her expertise. Knowing that Annan was a businessman rather than an inventor, she challenged his knowledge of mechanics at every turn. Then she delivered the crowning touch: a diary in which she had tracked the progress of her invention through a series of 1,867 entries. (Knight's preference for this number may have been related to the year 1867, which was when she began working at Columbia.) Claiming that the diary contained personal as well as professional information, she presented it for the eyes of patent examiners only.

In the end, these various tactics impressed the examiners, but they remained unsure on one point. A great deal of time had elapsed between Knight's conception of her machine and its successful production. Was it perhaps too much time? While the question may seem gratuitous to the uneducated, it did actually turn on a legitimate principle: inventions, then as now, were granted to the first to invent rather than the first to patent. Once an invention was publicized, however, the inventor had to file for a patent within one year. The purpose of this rule was to allow technology to go forward when people with good ideas proved to be too lazy to act on them.

Ironically, Knight's gender worked to her advantage here. The examiners considered her situation and decided that she was operating at a handicap, citing her "inexperience in business, as well as the embarrassment to which her sex [was] subjected." The commissioner of patents concurred and even went so far as to remark on her "most notable character."

In 1870, her patent secure, Knight went into a partnership with a businessman from Newton, Massachusetts, under terms that were highly favorable to her: $2,500 outright, 214 shares in their newly formed Eastern Paper Bag Company, and royalties on every paper bag sold until the total reached $25,000. From there, she moved into other realms of invention, and by the turn of the century was taking out patents on automobile valves, rotors, and engines. By then, she had been interviewed and held up as an inspiration to aspiring women many times over. As late as 1967, *Woman's Journal* was informing admirers of her home address—on the corner of Hollis and Charles, in Framingham, Massachusetts—whence they could go to pay their respects.

But in the end, for all her renown, Knight was unable to hang on to her wealth. Of the twenty-seven patents she received during her lifetime, she sold most to her employers for ready cash. When she died, in 1914, her personal estate was valued at $207.05—roughly enough today, in other words, to load up a few square-bottomed paper bags with groceries.

6
THE FAMILY ROOM

Night never falls completely in the American home. When the sun goes down, the TV screen illuminates it from within, parading a new set of products to replace the ones in the shadows . . .

Television

On May 20, 1924, the *New York Times* reported that AT&T had successfully sent still photographs over a telephone line. Asked by the reporter whether this accomplishment had brought the world one step closer to television—meaning images transmitted without the use of a projector—a company spokesman stated flatly, "Television will never happen." This must have come as news to Scotsman John Logie Baird, who had already transmitted the first televised images in the comfort of his studio. Then again, that's how things always went for Baird. A kind of John Henry of the communications industry, he was always winning the battle and losing the war.

Baird was born in 1888 and raised in Helensburgh, Scotland, near Glasgow. He showed an interest in theater at an early age and even as an adult could recite Goethe's *Faust* in its entirety. Certainly, he knew how to cut a dramatic profile.

After graduating from the Royal Technical College in Glasgow, he began to work as an engineer. The environments in which he worked were cold and drafty, and he often found himself in ill health. So it was that one day the landlady stumbled in on Baird—a lean young man with wire-rimmed glasses and a stylish shock of dirty blond hair on his brow—sitting in his chair, a wad of toilet paper wrapped around his feet. A man whose feet were forever cold, he had developed the habit of wrapping them in newspaper before donning socks and shoes. Toilet paper, he thought, might make for a more marketable material.

Baird pursued his "undersock" idea far enough to get fired from his engineering job of the moment, but it wasn't long before a new idea presented itself. Soon afterward, he ran into a childhood friend, Godfrey Harris, who convinced him to make a trip to Trinidad. Once there, Baird set himself up as a wholesaler. After three weeks, he had managed to sell no more than five pounds of safety pins, so he moved on to his next idea: a jam factory.

Baird commenced operations with two locals, known to history as Ram Roop and Tony, but as soon as they started to boil the jam, thousands upon thousands of insects descended upon the pot, astounded by their good luck. The team was soon overrun with cockroaches, spiders, and insects that defied categorization. When Baird came down with another fever and no interest whatsoever in his jam developed on the island, he returned to England and tried to sell his mango chutney and guava jelly to a nation very much accustomed to the finest jams in the world. In the end, he sold the lot for fifteen pounds, to a butcher who used it to stuff sausages.

Downtrodden and miserable, Baird started a soap company and took a room in a Bloomsbury boardinghouse, where a constant stream of transients tried to interest him in patent medicines. Fortunately, at this point another childhood friend reappeared to offer thoughtful advice. Guy Robertson, known to his friends as "Mephy" (short for Mephistopheles), convinced Baird to get out of Bloomsbury. Sick as ever, Baird sold his soap company for 200 pounds, and the two men then moved into a flat above an artificial flower shop on Queen's Arcade, in Hastings.

Baird's first project at Hastings was a glass razor, an idea that came to a swift end when he cut his face badly. Next he tried pneumatic shoes and "walked a hundred yards in a succession of drunken and uncontrollable lurches, followed by a few delighted urchins" before one of the balloon inserts in his shoes burst. Then one day, while rummaging around for a new project, he recalled some experiments he had conducted before the war and decided to resume where he left off, this time availing himself of the recent advances in vacuum-tube technology.

Inventors had been trying their hand at television for many years, using a simple device called a Nipkow disk. Devised in 1880 by a German named Paul Nipkow, this disk was little more than a circle with holes arranged in a spiral pattern. When it was spun, however, any light passing through it was dissected and carried by a selenium cell to a sister disk, which reversed the process and reproduced the particulars of the original light on a separate screen. Among Baird's competitors exploring the potential of this device were the Russian Vladimir Zworykin, the American Charles Jenkins, and, in fact, AT&T. In 1924, just after the company had stated that television "will never happen," Dr. Herbert E. Ives, an engineer at AT&T, began his own research on television.

Still, no one went about it the way Baird did. Improvising as best he could in his Hastings flat, he used, among other things, a tea chest, a "bull's-eye" lens from a local cycle shop, sealing wax, glue, surplus army wire, knitting needles, a hat box, a serrated biscuit tin, an ordinary lamp, and an electric fan. With this Rube Goldberg assemblage, the resourceful inventor defied all expectations and sent the silhouette of a Maltese cross a distance of two feet, probably in 1922.

Baird received a little money from investors for this achievement and used it to put an ad in the personals of the London *Times:* "Seeing by Wireless—Inventor of apparatus wishes to hear from someone who will assist (not financially) in making working models."

Apparently, the ad created the desired effect, because on April 3, 1924, an article by F. H. Robinson appeared in *Kinematograph Weekly*. One of the earliest printed references to a working television, "The Radio Kinema" described an experiment in which Robinson saw "the letter 'H,' and the fingers of my own hand reproduced . . . across the width of the laboratory."

Television work could be as perilous as the glass-razor business at times. As shoppers strolled along Queen's Arcade one day, they saw a flash of light emanate from Baird's flat, followed by a noisy fall. A crowd gathered, speculating about what had happened to the interloper in their midst. Someone finally gained entry into the flat and found Baird unconscious, felled by a surge of 2,000 volts.

When an article came out announcing, "Serious Explosion in Hastings Laboratory," Baird's landlord,

Mr. Twigg, decided that the inventor was a liability and told him to pack his bags. Baird returned to London, where again he met with reservations. When he was invited to see an editor at a London paper, one newspaperman saw him arrive at the office and cried out: "Watch him; he may have a razor on him!"

Baird's star began to rise again when Gordon Selfridge offered him 20 pounds a week to give three shows a day in his London department store. Spanning three weeks in April 1925, the Selfridge demonstrations drew long queues of spectators. Most of the visitors simply looked into a funnel and saw the outlines of objects that stood a few yards away—although one woman reportedly worried that the machinery would damage the racks of clothing nearby.

Six months after the Selfridge demonstrations, Baird used his new "flying-spot" mechanism—a modification in the way the light passed through the Nipkow disk—to televise what most experts agree was the first image with a full range of halftones. He chose the leering face of Stukey Bill, a ventriloquist's dummy, to be television's first star. Unable to contain his excitement, he immediately buttonholed a young office boy, William Taynton, who worked in the building. Moments, later, Taynton became the first living face seen on a television screen.

On January 27, 1926, Baird unveiled his half-tone television for more than forty members of the Royal Institution. In full evening dress, a pride of eminent scientists and their wives climbed three flights of narrow stone stairs, then stood in a cramped, drafty passage while groups of six were led into two tiny attic rooms. Here, a picture made up of thirty lines, seven inches high by three inches wide, traveled through a wall to a comparably shaped receiver on the other side.

One hirsute gentleman peered a little too close to the fascinating device and found his long white beard being sucked into it. Fortunately, he was rescued in time, and when he approached the camera again a few minutes later, he had the pleasure of seeing his face—and most of his whiskers—transmitted live onto a television screen.

The press responded enthusiastically to the Royal Institution demonstration, publishing accounts entitled "Magic in a Garret" and the "Young Scotsman's Magic Eye." Baird took the bit and ran with it. In 1926, he invented an early prototype of fiber optics. That same year, he obtained the first television license, with the call letters 2TV, and began selling simple television sets through his own magazine, which bore the appropriate title of *Television*. On January 26, 1927, he filed for a patent on a magnetic recording system capable of recording a single frame or of playing back a recording at high speed—the ancestor of videotape, complete with recording and playback heads.

On April 7, 1927, AT&T did a complete about-face, demonstrating its own version of the idiot box with the slogan "Television At Last!" Not one to be caught short, Baird promptly sent an image from London to Glasgow, another from London to America, and, finally, one from England to an ocean liner in the middle of the Atlantic. Needless to say, a photograph materializing on a crude screen, set up in the telegraph room belowdecks, caused considerable stir among the passengers. But it was on the following day, back in London, that Baird worked up his most unlikely experiment of all. On February 10, 1928, he contrived to procure a human eye from a London surgeon. "As soon as I was given the eye," Baird recalled,

> I hurried in a taxicab to the laboratory. Within a few minutes I had the eye in the machine. Then I turned on the current and the waves carrying television were broadcast from the aerial. The essential image passed through the eye within a half an hour after

the operation. On the following day the sensitiveness of the eye's visual nerve was gone. The optic was dead. . . . Nothing was gained from the experiment. It was gruesome and a waste of time.

Indeed—and Baird could no longer afford to waste any time. In July, he transmitted color television in a closed circuit in his lab. "We found that strawberries came out particularly well," he remarked good-humoredly, "and they were popular with the staff." (Unfortunately, the British papers didn't cover this achievement.) A career peak came on July 28, 1929, when Baird televised a theater performance from the Coliseum, in London's Charing Cross theater district, to those who had bought his sets. But it was all downhill from there.

When the BBC television station 2LO went on the air on September 30, 1929, television was well on its way to widespread acceptance. Baird himself was never associated with the station, however, thanks to a rift that formed between him and John Reith, founder of the BBC and a friend of Baird's from childhood. The discord only grew worse as time wore on, and eventually Baird was forced to operate an independent station out of a building at 133 Long Acres, where the admiralty at Whitehall complained of interference on their radio frequencies.

In point of fact, even the BBC was behind the times by this time. While AT&T and Baird had been battling for the television prize, RCA president David Sarnoff had quietly hired Vladimir Zworykin, recently emigrated from Saint Petersburg, to develop an all-electronic television system. By 1933, Zworykin was able to report success. In the subsequent years, a brilliant boy wonder from Utah, Philo T. Farnsworth, laid claim to the same achievement. Electronic television was far superior to the mechanical systems Baird had developed, and RCA, having cornered the rights to both of the electronic versions then existing, soon emerged as the preeminent power in American television.

In his final years, Baird continued to work on various technologies, and he even operated a secret lab inside the old Crystal Palace in London during World War II. He took up yoga in an age when Eastern philosophies were still considered exotic, and shortly before his death, in 1945, he could be found musing aloud to his assistants about the spiritual potential of television.

The *spiritual potential* of television?

Too bad he lost the race!

The Exercise Machine

The Nautilus, which set the fitness craze in motion when it was introduced at a Los Angeles weight-lifting conference in 1970, was specifically designed to be intense. Rather than moving gradually toward peak performance, the Nautilus user was expected to begin at maximum effort and then maintain that level for as long as possible. Some experts have quibbled with this approach, and others have likened it to torture, but no one has ever argued that the Nautilus principle didn't match the personality of its inventor.

Quite simply, Arthur Jones is among the darkest and crankiest people you're ever likely to meet. A lifelong smoker and a perennial bride-groom, he rarely utters a sentence that can be printed in a family magazine. He believes that America is long overdue for a revolution, and, if pressed, will insist that Hitler did some pretty good things. Asked to define his credo, he says bluntly, "The two things that motivate people are terror and greed."

Such sentiments may stem from the fact that Jones's upbringing resembles something out of *Huckleberry Finn*. He was born in Arkansas and lived there until the early 1920s, when his family relocated to Oklahoma. The young Arthur had little patience for the pleasures of the hearth. He ran away from home for the first time at the age of eight, and made off for good just in time for puberty. Freed from parental control, he married early and often. He was a still a minor when he took his first wife, a nineteen-year-old. Two years later, it was the bride who was jailbait. His fifth wife, fetched when he was in his fifties, had yet to pass the nineteen-year mark when she arrived at the altar.

Between romantic entanglements, Jones has lived a life of almost constant adventure. A few of his exploits, such as his alleged tours of duty as a soldier of fortune, remain too hot to touch even today. "I've been in several wars," he hedges, "but I'd prefer not to talk about them. If you were ever in a war, you would understand." Most, however, are a source of unmitigated pride.

An expedition to Africa in 1956, for example, found Jones capturing adult crocodiles in the bush and living to tell the tale in glowing colors. "In my biased opinion," he wrote of this episode, "I believe that my trip to Caprivi was the greatest adventure in history." Of course, the success of Nautilus, achieved after two decades of painstaking research, became its own kind of adventure, as the sudden influx of millions of dollars gave new strength to his personal obsessions. A pilot since his early adulthood, he soon owned a minor squadron of jets and a private runway on the grounds of his Florida home for the importation of exotic animals. He generally sold these animals for profit, but a good many he kept for himself. At peak, his menagerie included forty crocodiles, dozens of snakes, tarantulas, and the largest privately owned herd of elephants on earth.

During his salad days, Jones also took on the ambitious scheme of producing his own cable television show. With his personal friend G. Gordon Liddy as host and a shark-filled tank for a backdrop, things looked promising. But apparently even Liddy was too soft for Jones. After the infamous ex-plumber granted the head of the American Nazi Party the right to his opinion—that time-honored liberal entitlement—during one interview, Jones scuttled the show. "You didn't get your reputation," he told Liddy, "by being a left-wing pansy."

In 1986, Jones sold his interest in the Nautilus business—without having paid a nickel in taxes—and sank his remaining fortune into MedX, a series of machines for the diagnosis and rehabilitation of back and knee problems. Between superlatives, he characterizes this latest invention as being "up there with penicillin, X rays, or the fucking hypodermic needle."

That the scientific community has been slow to agree doesn't faze Jones a bit. On the contrary, he's apt to approach scientists with the same bravura that he once brought to the capturing of crocodiles. Whenever an unsuspecting professor scoffs at his presentation, he says, he simply lodges a finger behind the man's Adam's apple, grabs him by the family jewels and beats the living daylights out of him.

"Do it right," Jones points out, "and they'll never say anything. Don't hesitate to use force. In the end, all problems are solved by force. But do a good job. Superior force is not enough. Unhesitating willingness is not enough. You must indicate that in fact you are *anxious* to use force, and that if they don't give you an excuse to use it, you will invent one."

If this seems like sound advice, you might want to bone up on your Nautilus first.

United States Patent [19]

Jones

[11] **3,998,454**

[45] Dec. 21, 1976

[54] **FORCE RECEIVING EXERCISING MEMBER**

[76] Inventor: Arthur A. Jones, Lake Helen, Fla. 32744

[22] Filed: Sept. 17, 1974

[21] Appl. No.: 506,795

Related U.S. Application Data

[60] Division of Ser. No. 360,590, May 15, 1973, Pat. No. 3,858,873, which is a continuation of Ser. No. 172,478, Aug. 17, 1971, abandoned.

[52] U.S. Cl. 272/117
[51] Int. Cl.² A63B 21/00
[58] Field of Search .... 272/51, 57 D, 71, 73, 272/117, 118, 93, 178, 144, 116; 128/25 D

[56] **References Cited**

UNITED STATES PATENTS

684,688	10/1901	Herz 272/81
3,475,024	10/1969	Lewis 272/81
3,640,527	2/1972	Proctor 272/81

FOREIGN PATENTS OR APPLICATIONS

495,945	2/1938	United Kingdom 272/73
215,774	7/1968	U.S.S.R. 272/81

Primary Examiner—Richard C. Pinkham
Assistant Examiner—William R. Browne
Attorney, Agent, or Firm—Karl W. Flocks

[57] **ABSTRACT**

A force applying member for use in an exercising machine wherein the member includes a pair of generally L-shaped levers. A first leg of one L-shaped lever is parallel to a first leg of the other L-shaped lever and a second leg of one L-shaped lever extends generally parallel to a second leg of the other L-shaped lever. A member is attached to the first legs for receiving a force applied by a user. The latter member is adjacent the intersection of each of the first and second legs of each L-shaped lever. There is an angled portion on each of the second legs at an end away from the respective intersections of said first and second legs and a transversely extending stabilizing bar connecting the angled portions and having a clearance therebetween. Each first leg has a rotatable hub member integrally extending from one end thereof.

3 Claims, 14 Drawing Figures

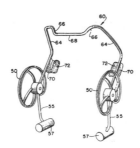

United States Patent [19]

Jones

[11] **3,858,873**

[45] Jan. 7, 1975

[54] **WEIGHT LIFTING EXERCISING DEVICES**

[76] Inventor: Arthur A. Jones, P.O. Box 1783, Swartz Creek, Mich. 32720

[22] Filed: May 15, 1973

[21] Appl. No.: 360,590

Related U.S. Application Data

[63] Continuation of Ser. No. 172,478, Aug. 17, 1971, abandoned.

[52] U.S. Cl. 272/58, 272/81, 272/DIG. 4
[51] Int. Cl. A63B 21/00
[58] Field of Search .... 272/81, 83 R, 82, 80, 79 R, 272/DIG. 4

[56] **References Cited**

UNITED STATES PATENTS

3,116,062	12/1963	Zinkin 272/81

FOREIGN PATENTS OR APPLICATIONS

215,774	1964	U.S.S.R. 272/81
12,636	5/1910	France 272/81

Primary Examiner—Richard C. Pinkham
Assistant Examiner—William R. Browne
Attorney, Agent, or Firm—Karl W. Flocks

[57] **ABSTRACT**

An apparatus for the development of body parts and muscles effecting the movement of a user's body parts. The apparatus includes a frame on which is pivotally mounted a force applying member against which a body part may be placed and urged for purposes of developing said body part and muscles. The user is positioned on the front side of the frame during an exercise program. The apparatus also includes a weight pivotally mounted on the rear side of the frame and a force resolving spiral means rigidly connected to the force applying member for rotation therewith whereby the pull on the weight mass is continuously varied over the full range of rotation of said force applying member to produce a varying resistance force in direct opposition to the force applied to the force applying member by a body part and thereby to stress and develop said body part and muscles. The resolving spiral is in the form of one or more pulleys which continuously resolves the force of the weight mass to thereby provide optimum stress for developing the muscles. The effort of continuously exposing the body part undergoing an exercise to a varying force over the full range of movement will cause the muscle effecting movement of said body part to be subjected to optimum development conditions.

13 Claims, 14 Drawing Figures

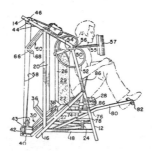

Solitaire

C redit for the invention of solitaire depends on who is telling the story. Some say it was developed by a man named by Pélisson, who, having been imprisoned by Louis XIV, had more than enough time to invent whatever caught his fancy, provided it required no more than a deck of cards. Others insist it was a mathematician who grew bored during his many long coach journeys.

The timing of both legends is about right. When playing cards were first introduced into Europe, in the late 1300s (before the Gypsies, who supposedly brought them, had arrived, and before tarot cards, upon which they were supposedly based, had been invented), they were hand-painted luxury items, with decks that varied widely from country to country and year to year.

It took the invention of the printing press, circa 1440, for cards to become the plaything of the masses, and so to become standardized into the familiar fifty-two-card deck of today. By 1790, the daughters of the daughters of the original European Gypsies—who were very much part of the masses—began to arrange their tarot cards in a "layout" style. At about the same time, the first solitaire games based on the positioning of the cards appeared as well. (Among the first variations were Grandfather's Patience and Sir Tommy, the latter of which is played today under the name Old Patience.) In a sense, then, there is a connection to be drawn between Gypsies and solitaire. One could even go beyond historical parallels and make a case for a thematic similarity. After all, where the fortune teller divines the future of a paying listener, the solitaire player plays chicken with *statistics*.

Solitaire was introduced into the United States in due time, of course, and has been played here ever since. But Americans will rarely leave a simple thing be when they can complicate it with machinery. Thus the sight of one person alone playing a game of cards has been replaced in recent years by the sight of a solitary figure playing against a computer. Not even a fortune teller could have predicted how this came about.

In 1958, Willy Higinbotham was working at the Brookhaven National Laboratory, a government-supported research facility in Upton, New York.

Brookhaven was devoted to nuclear energy rather than nuclear bombs, but, the fear of global annihilation being what it was, the laboratory had begun to organize tours that emphasized the safety and importance of "Atoms for Peace."

Higinbotham had no problem with that. His complaint was that the displays consisted largely of photographs and dull-looking equipment. What the thing wanted was a little pizzazz, so he began casting about for a way to provide it. His department owned a crude computer that, when combined with an oscilloscope, a few capacitors, potentiometers, and unspecified household items, displayed the trajectories of missiles or bullets. (Why a nuclear energy lab was tracking missiles is hard to say, but that's how the story goes.) For Higinbotham, who had designed timing devices for the Manhattan Project, it was mere child's play to convert the imagery on the screen from bullets to a tennis ball and court.

By today's standards, the first video game was laughably crude. Using conventional vacuum tubes, a time-sharing circuit, a few transistors (a new technology in those days), and a resistor for simulating wind drag, he created a court, a net, and a ball, all of which appeared in rapid alternation. Still, visitors were transfixed. They pushed a button to set the ball in play, then turned a knob to aim it. Then they did it again. As one scientist later recalled, "People would stand on line for hours to play it."

The game remained part of the tour for the next two years and went through several elaborations to keep up the public interested. At one point, the court was relocated to outer space, with "astronauts" playing at gravity levels equivalent those on Mars or Jupiter. Eventually, though, the enthusiasm for computer tennis waned, and the machine was dismantled. Higinbotham, whose real job was to design radiation detectors, never even bothered to patent the idea, though he lived to regret it. In 1972, Noland Bushnell stole Higinbotham's thunder with the video game Pong—itself a simulated tennis game—and the video-game industry began in earnest.

The parallels between the old "carbon-based" solitaire and the video game are worth savoring. Where the first playing cards turned the military imagery of swords and kings into a pleasant pastime, the first video game turned missiles into tennis balls. And one can only imagine what Higinbotham, who had witnessed some of the first detonations at Los Alamos, would think of the mass destruction played out on video screens today.

Watching the mushroom cloud rise, did he perhaps feel a solitude so great, so *imprisoning,* that the only rational response was to formulate a game?

If not, it nevertheless became the case for the rest of us. Throughout the ages, humans have engaged in magical thinking, bargaining for their destiny in the random signs around them. In an age of royalty, orphans rearranged their majesties while waiting for the next calamity. In the age of nuclear warheads—which is still with us, mind you—we bury ourselves, individually and endlessly, in a world of explosions, keeping track of our progress through statistical ratings.

Of course, the screw always turns an extra time, and today actual wars are being conducted as if they *were* video games, with cameras strapped on the warheads and sailors firing them from aircraft-carrier decks hundreds of miles from ground zero. Somehow, it's hard to imagine the next version of solitaire coming out of *that.*

7
THE BEDROOM

The bedroom, unlike the other rooms in your house, is based on the idea of subtraction. All of the objects used or accrued throughout the day are surrendered here, falling away one by one . . .

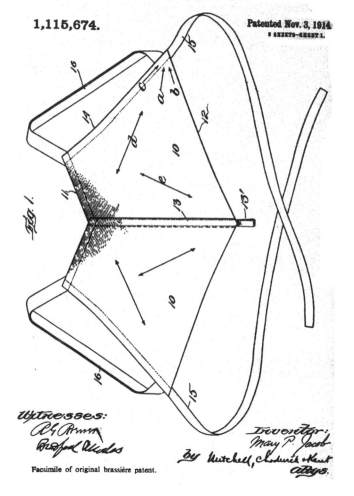

1,115,674.

Fig. 1.

Witnesses:

Inventor:
Mary P. Jacob
by Mitchell, Chadwick & Kent
attys.

Facsimile of original brassière patent.

The Brassiere

A mong the most enduring images from the 1960s is that of women burning their bras. The gesture was meant as an act of liberation, of course, and the bra was decried as a stricture imposed by men. We can count this as one more example of history proceeding by amnesia, however, because the brassiere was originally designed to liberate, and its inventor was among the most freewheeling women of her day.

In her autobiography, *The Passionate Years,* Caresse Crosby insists that her parents were "never rich," although it's difficult to keep this statement in focus for very long. The child of a robber baron in Edith Wharton's New York (she was born in 1898), Crosby knew J. P. Morgan well enough to address him as "Uncle Jack," and her early life was an unending stream of gala events: one summer, she attended as many as three social functions a day, with regroupings between at Delmonico's. Fame was never far from her reach, either. As a teenager, she was invited to the circus by songwriter Cole Porter and managed the honor of becoming the first Girl Scout of America.

In any event, Crosby was still going by the name Polly Jacobs in 1914 when her silver-spoon existence took a quiet but definitive turn. Society women of her day were expected to don a stable's worth of undergarments, which in turn required a good deal of calculation before the looking glass. One day Jacobs, deep into this very brand of strategizing, was refurbishing a wreath of silk roses that matched the rose-garlanded dress marked for that evening's soiree. Then she remembered how the eyelet embroidery in her corset cover tended to poke through the roses around her breasts.

"I'm not going to wear that thing tonight," she told Marie, her lady-in-waiting.

"But Mademoiselle cannot go without a *soutien-gorge*," Marie replied evenly.

To read the inventor's account, the solution came in an instant. "Bring me two of my pocket handkerchiefs," she commanded, "and some pink ribbon ... and bring the needle and thread and some pins into my bedroom."

Sitting before her mirror, Jacobs pinned the handkerchiefs together, then stitched the pink ribbons to the two opposite ends. Marie pulled the ribbons taut and tied a knot behind. The result was not a bra as we understand the term today. At the time, the desired effect was "to flatten down one's chest as much as possible," Jacobs later explained, "so the truth that virgins had breasts should not be suspected."

But it was a bra nonetheless. That night, Jacobs floated through the party looking especially "fresh and supple." Afterward, a gaggle of jealous friends gathered round to discover her secret, and the swains of Manhattan were soon enjoying a spike of virgins among the glitterati.

Despite the sudden popularity of her brassiere, Jacobs did not think to market it until a stranger from Boston wrote asking for one—and even then she went about it very much on the sly. She secretly contacted a patent lawyer, who drew up a set of diagrams and asked for a fee of $50. Having been raised never to touch money, she could offer only five on account, with the promise of more when the patent was awarded. Then she went off to Europe, and the lawyer went to work. When she returned, she found a patent for the "backless brassiere" waiting to greet her.

At this point, Jacobs went into business proper. Stretching the rules of the Social Register to the breaking point, she borrowed $100 from an elderly friend, made her lady-in-waiting Marie a partner, hired two underage Italian girls, and started a sweatshop—all without anyone in society circles hearing a word. When the sweatshop turned out a few hundred units, Jacobs tried without success to sell her patent to some big department stores of the time. Finally, she caught the attention of Johnny Fields, who worked for the Warner Brothers Corset Company. Fields bought her patent for $1,500, and that, in effect, was the end of her career as an inventor.

It was hardly the end of her life, though. The backless brassiere caught on like wildfire in the 1920s (arguably creating the "flapper" look), and Jacobs lit up accordingly. After a failed marriage to the rich but bland Richard Peabody, she ran off to Europe with a banker-poet named Harry Crosby, whom she loved at first sight and eventually married.

The Crosbys made for quite a pair. They dabbled in opium and followed a vaguely Incan brand of sun worship. One night, Jacobs made her entrance at a Quatre Arts Ball by riding down the Champs-Elysées on a baby elephant, while Harry held court on the ground wearing a necklace of dead pigeons.

As the night wore on, Mrs. Crosby made a topless appearance onstage, and Harry found himself in a bathtub with three women he had never met before.

Polly Crosby, who was already on her third surname, shed her first name thanks to Harry's infatuation with acrostic poetry—a parlor game in which the first letters of each line in a poem, when read vertically, spell out the name of a loved one. Casting around for both a poem and a new moniker for Polly, they arrived at "Caresse." At the time, this was a scandalous name, and it was quickly declared the equivalent of "undressing in public." But since Polly had already done that, Caresse could not be bothered much by the criticism. She defiantly kept the name, French "e" and all.

The wild life—or something hidden beneath it— eventually became too much for Harry and drove him to suicide. Caresse, on the other hand, maintained her sense of adventure to the end. "I have learned," she wrote in her autobiography, "that personal life is the individual's only means of expression in a cosmos forever mysterious. It is the right to this life itself that must be made secure for the unborn citizens of a challenging universe. Like Harry, I believe there can be no compromise. The answer to the challenge is always 'Yes.'"

It would be wrong to accuse the Crosbys of squandering everything, however. Between blow-outs, they made themselves useful by starting the Black Sun Press, which introduced such writers as D. H. Lawrence, James Joyce, and Ezra Pound to the American reading public. It might even be argued that the invention of the brassiere, rather than imprisoning women, led Caresse Crosby to the very cultural machinery that eventually allowed feminism to flourish. It was only a few transpositions, after all, from the publication of Molly Bloom's erotic soliloquy in Joyce's *Ulysses* to public demonstrations of feminine lust.

"Yes" indeed.

Shatterproof Glasses

he invention of plastic lenses says a lot about the nature of genius and the fruits it may bring. Unfortunately, none of it does much to celebrate the genius in question.

Robert K. Graham grew up in northern Michigan, the son of a dentist. Even as a child, Graham had unusual preoccupations. He noticed, for example, that the most eminent people in his neighborhood were terribly remiss when it came to procreation. The local doctor and banker each had only one child. The richest man in town had no offspring at all. Sir Francis Bacon had once made a similar observation and concluded that bachelorhood spurred bachelors to excel. Graham turned that proposition around and began cultivating a fear that the stupid would soon rule the earth.

The prospect of an idiot planet stayed with Graham even after he came of age and began working as a medical salesman. Eventually, it bothered him so much that he wrote a book called *The Future of Man,* a polemic on the virtues of eugenics. Graham showed the manuscript of this book to Hermann J. Muller, who had fetched a Nobel Prize for his work on radiation and genetic mutations. Not coincidentally, Muller was also a eugenicist—and apparently one after Ayn Rand's heart at that.

"There is no permanent status quo in nature," Muller declared in 1935, "all is a process of adjustment and readjustment, or else eventual failure. But man is the first being yet evolved on earth which has the power to note this changefulness, and, if he will, to turn it to his own advantage, to work out genetic methods, eugenic ideas, yes, to invent new characteristics, organs, and biological systems that will work out to further the interests, the happiness, the glory of the God-like being whose meager foreshadowings we the present ailing creatures are."

Whether or not Graham shared Muller's interest in inventing new organs (for what in 1935 was beginning to look like an awfully overorganized human race), their three-day meeting in Graham's Pasadena home marked the beginning of a lifelong friendship. But camaraderie alone was not enough. What remained was for Graham to apply the ideas in his book—and for that he had to expend some effort at being a genius himself.

Exactly when and how Graham invented plastic lenses is generally unremarked in the literature of the field. Did it come about by accident? Did he have a "eureka" moment? In a logical world, he would have started with the same event that windshield-wiper man Robert Kearns did—that is, with a champagne cork in the eye. But the world is not a logical place, and the invention of shatterproof glasses does not come with a creation myth attached. The best that can be said is that in 1947 Graham organized a company in Burbank, California, called Armorlite, which produced plastic lenses—known in the business as CR-39 lenses—until the late 1970s, when Graham sold out to 3M. (Armorlite subsequently broke free of 3M and paired off with the lens manufacturer Signet to form Signet Armorlite.)

Graham walked away from the sale not only rich but inspired. For about three years, he had been quietly organizing a sperm bank—initially in the pumphouse on his California property—that was founded on a simple premise. Using sperm donated by Nobel Prize winners, bright married women with sterile husbands could give birth to potentially brilliant children. (Graham's sperm bank, interestingly enough, relied on the technology developed by Clarence Birdseye: sperm, like food, has to be frozen quickly in order to prevent crystals from forming.)

The donors themselves were expected to remain anonymous—a decision that led to some bizarre circumlocutions. As Graham explained it, "Paul [Smith] makes our collections from donors. They wish it to be anonymous, so when Paul appears on television, he always wears a surgeon's mask so he won't be recognized. He also makes some deliveries of the germinal material, and the husband doesn't want Paul to be recognized, either. There are a lot of delicate feelings involved in this whole project. So we have to maintain absolute anonymity."

The Hermann Muller Repository for Germinal Choice, as Graham proudly dubbed his endeavor, made the papers in March 1,1980, when William Shockley, co-inventor of the transistor, broke the anonymity rule and announced that he had become a donor. Shockley, himself a known proponent of eugenics, offered a rationale to the *New York Times* that was as circuitous as it was defensive: "I feel this whole area of subject matter is being woefully neglected and that the humanitarian instincts that have led to this neglect are a form of humanitarianism that has gone berserk."

Predictably, the humanitarians went even more berserk. A *Times* editorialist beheld a photo of Shockley, bespectacled and bald, and compared him unfavorably to Elvis Presley. Muller's widow, Thea Muller, publicly protested the use of her husband's name in connection with the sperm bank, and Graham was compelled to shorten the name to the simpler Repository for Germinal Choice.

Asked in an interview why Ms. Muller objected even though her husband (who died in 1967) had been a eugenicist, Graham later suggested that she had been upset by the character of some of the recipients. But when Ms. Muller made her complaint, on March 4, 1980, those characters were yet to come.

In July 1982, Joyce Kowalski of Phoenix, Arizona—one of the first recipients of donated Nobel sperm—was found to be an ex-convict, and to have lost custody of two children from a previous marriage after her current husband was accused of child abuse. And that was just for starters. Kowalski and her husband, Jack, had also been convicted in May 1978 of using the records of dead children—of all things!—to secure credit cards and bank loans. Four days later, while reporters were still trying to break the interview embargo imposed on the Kowalskis by the *National Enquirer,* the second repository mother, Afton Blake of Los Angeles, revealed that she was unmarried—a direct violation of her contract.

Graham kept up a game face for a time. On September 23, 1982, he appeared at the New York Academy of Sciences to whip up support for his project. By then, some twenty "eminent men" had donated their sperm, and as many as four hundred women had signed up to become recipients. Defending the company's position to a *Times* reporter, research officer Paul Smith insisted, "We want the entire human race to be a master race."

But by 1984, Graham was visibly on the defensive. Noting that women responded poorly to the visage of elderly donors—as most Nobelists were— he began to select his donors from the ranks of potential Nobelists and, for good measure, Olympic-level athletes. He also relaxed at least one of the standards for the recipients of the sperm: they no longer had to be exceptionally bright.

In effect, the Repository for Germinal Choice had become like any other sperm bank. It appears to have been moderately successful in this capacity: by the close of 1991, the bank was responsible for some two hundred births. Meanwhile, Graham, when not making improvements on his lenses, began cultivating the high art of the pitchman. As of this writing, one can find a Web site devoted to a video called *Speaking with Dr. Graham,* the promotional material for which—delivered here verbatim—speaks for itself:

Speaking With Dr. Graham (improve your sperm count)

Seemingly intractable problems with infertility, especially in the male, are frequently not as complicated as one might assume. Often, a male can actually be fertile, but a chronically low sperm count will inhibit successful pregnancy.

To alleviate this problem, Dr. Robert Graham, the founder and operator of the genius sperm bank, after years of experience in the business of providing the healthy sperm to exceptionally bright women for use in fathering their children, came up with a simple regimen that will dramatically increase the sperm count and health of such for those so afflicted.

A by-product of this development is a substantial increase in sexual frequency and ejaculate volume which also increases sexual satisfaction.

100% natural and anyone can benefit.

Find out how by ordering the videotape, Speaking With Dr. Graham. $30.00 ea., postpaid, while supplies last. (Not recommended for those under 18 yrs. of age.)

Send check or money order to . . .

Of course, the wearing of shatterproof glasses while viewing is recommended. Just in case.

The Condom

What's in a name?

The Paduan professor of anatomy Gabriel Fallopius is best remembered for lending his name to the fallopian tubes, but he was also the first European to address the problem of male contraceptives. In the mid-1500s, Fallopius devised a cloth covering that was placed over the glans penis and tucked under the foreskin—not to prevent pregnancy, which was considered the woman's problem, but to stanch the spread of venereal disease. Fallopius claimed to have experimented on eleven hundred men in perfecting this device, although exactly how he conducted his trials remains untold.

As tradition would have it, the modern condom was invented by the personal physician of Charles II of England, who went by the felicitous name of Dr. Condom. But traditions have a funny way of gathering around thin air. Indeed, according to

William E. Kruck, who took the time to write a fifty-seven-page article titled "Looking for Dr. Condom," the legend of the personal physician has no basis in fact at all.

The word *condom* was first used to describe a contraceptive device in 1705, when the Duke of Argyll is recorded as having entered the Scottish Parliament with a device made of animal intestine, which he called a "quondam." The following year, Lord Bellhaven wrote a poem containing the word *Condum*—the first example of the word in print. The word became officially equated with a person in London in 1708, when the play *Almonds for Parrots* included a character named Condun, who was said to have invented the device. By 1724, the whole business had already become something of an urban legend—so much so that one Dr. Daniel Turner could only speculate that "Dr. C——n" was *probably* the inventor.

Over the years, these bits and pieces of documentation have done much to establish the existence of a Dr. Condom in the popular mind. Unfortunately, countless investigations by etymologists (who pursue such mysteries for a living) have unearthed no evidence whatsoever of a Dr. Condom in any of the plausible spellings. He does not show up in the court of King Charles II, and he certainly does not appear as the Earl of Condom, as some sources have suggested. In fact, Condom was not even a known English name in the seventeenth century, and it remains uncommon today.

Of course, it is conceivable that the inventor used a false name out of embarrassment, but even if this were true, one would still have to explain why he chose the particular pseudonym he did. What special advantage did the name Dr. Condom have over, say, Dr. Raincoat?

According to Kruck, there are three working explanations for the derivation of this elusive word: the name of a French town, the Latin word *condus,* and a Persian pun. Having set up these ducks, he proceeds to knock them down one by one.

The common pronunciation of the name of the French town Condom, Kruck points out, rhymed then as it does now with the English word *hobo,* and therefore would never have arrived in English as *quondam,* especially by the vernacular routes such a word was likely to take. What's more, Casanova (who pursued another kind of mystery for a living) wrote in his memoirs of *redingotes anglaises,* or "English riding coats"—and when did a lover ever credit the English when a French explanation would do?

The Latin-derivation theory, put forward by a

German doctor named Arnold Meyerhof—writing under the pseudonym Hans Ferdy—fares no better under Kruck's scrutiny. As Ferdy explains it, the word evolved from the Latin *condus*, which translates as "that which secures, preserves, guards [something]" and takes the accusative *condum*. The problem here is that *condus* is a neologism, or coined word, based on the Latin *condere*, which translates as "to lay up, found, to put together." Nor is the meaning of the word the only obstacle: *condus* appears only twice in all of Latin literature, and both times at a great remove from the seventeenth century. It first appears in 191 B.C., when Plautus uses it in a play, then lays low for some two hundred years until Ausonius grabs hold of it for his Epistle to Paulinus. After that, *condus* vanishes until Ferdy himself comes up with his theory.

We can thank another German doctor, Paul Richter, for the theory that *condom* derives from a Persian pun. As Richter had it, the Persian for an earthen grain-storage pot, *kondü* or *kendü,* was borrowed into Greek and then into Latin by a medieval scholar eager to make witty remarks about male sheaths. Richter, however, stood entirely alone in this explanation, not only because the scholar he described was entirely hypothetical but more damningly because the word *condom* entered the language in the seventeenth century, making any concatenations from the Dark Ages nigh on impossible.

And so we finish our tour of the House of Invention in, if you'll excuse the expression, a shroud of uncertainty. This is only appropriate, of course, since the condom itself suffers from an abiding lack of charisma and is best administered in darkness. But for the sake of completeness, we might leave off with a final thought.

Dr. Condom, one might say, is the man who *would* have existed had generations of lotharios never used the device that bears his name. He exists in every man, everywhere, waiting for the day when he may announce his arrival at last.

Suggested Reading

Berton, Pierre. *Niagara: A History of the Falls*. New York: Kodansha America, 1997.

Bunch, Bryan H., and Alexander Hellemans. *The Timetables of Technology*. New York: Simon & Schuster, 1993.

Crosby, Caresse. *The Passionate Years*. New York: Ecco Press, 1972.

Davis, Andrew McFarland. *Currency and Banking in the Province of the Massachusetts Bay*. New York: Augustus Kelly, 1970.

Flatow, Ira. *They All Laughed . . . : From Light Bulbs to Lasers—the Fascinating Stories Behind the Great Inventions That Have Changed Our Lives*. New York: HarperCollins, 1992.

Freeman, Allyn, and Bob Golden. *Why Didn't I Think of That? Bizarre Origins of Ingenious Inventions We Couldn't Live Without*. New York: John Wiley & Sons, 1997.

Fucini, Joseph J., and Suzy Fucini. *Entrepreneurs: The*

Suggested Reading

Men and Women Behind Famous Brand Names and How They Made It. Boston: G. K. Hall & Co., 1985.

Giscard d'Estaing, Valerie-Anne, and Mark Young, eds. *Inventions and Discoveries 1993: What's Happened, What's Coming, What's That?* New York: Facts on File, 1993.

Hobbs, Alfred C. *The Study of Locks and the Mechanism of Opening Them.* Bridgeport, Conn.: Gould & Stiles, 1889.

Kronstain, Jennifer. "Clarence Birdseye: Carving a Niche in the Food Industry." *Entrepreneurial Edge Online,* Edward Lowe Foundation, 1998.

Kruck, William E. "Looking for Dr. Condom." *Publication of the American Dialect Society 66*:1–58.

Lanza, Joseph. *Elevator Music.* New York: St. Martin's Press, 1994.

Lindsay, David. *Madness in the Making: The Triumphant Rise and Untimely Fall of America's Show Inventors.* New York: Kodansha America, 1997.

Macdonald, Anne L. *Feminine Ingenuity: Woman and Invention in America.* New York: Ballantine Books, 1992.

Manning, William. *Recollections of Robert-Houdin.* Chicago: H. J. Burlingame, 1898.

Music in Industry. Chicago: Industrial Recreation Association, 1944.

Powell, Horace B. *The Original Has This Signature—*

W. K. Kellogg. Englewood Cliffs, NJ: Prentice-Hall, 1956.

Schwarz, Frederic D. "The Patriarch of Pong." *American Heritage of Invention and Technology* Fall 1990; 64.

Seabrook, John. "The Flash of Genius." *The New Yorker* January 13, 1993: 38–52.

Sonneborn, R. M. "Muller, Crusader for Human Betterment." *Science* 162 (1968): 772–76.

Squier, George Owen. *Telling the World.* Baltimore: William & Wilkins, 1933.

Stone, John Stone. "The Practical Aspects of the Propagation of High-Frequency Electric Waves Along Wires." *The Journal of the Franklin Institute* 174.4 (October 1912): 353–374.

Vare, Ethlie Ann, and Greg Ptacek. *Mothers of Inventions: From the Bra to the Bomb, Forgotten Women and Their Unforgettable Ideas.* New York: William Morrow & Co. 1987.